FIELD GUIDE TO THE

wildlife

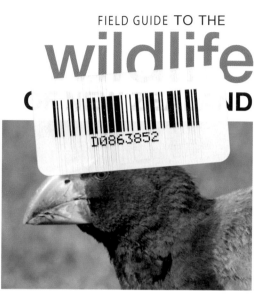

Birds • Mammals • Amphibians • Reptiles
Insects • Seashore • Trees • Shrubs •
Vines • Ferns • Alpine Plants • Herbs •
Grasses • Fungi and Lichens

JULIAN FITTER

BLOOMSBURY WILDLIFE
LONDON • OXFORD • NEW YORK • NEW DELHI • SYDNEY

To my father, Richard, who thought I was mad to try, but who would have loved the result, and for Jayne, who made it possible and enjoyable.

Above: A stand of Nikau Palms *Rhopalostylis sapida*.
Title page: Takahe *Poryphio hochstetteri*.

Main cover image © John Cancalosi/naturepl.com
Front cover (clockwise from top left) Julian Fitter, Julian Fitter, John Marris/ Natural Sciences Image Library, Richard Chambers, Tui De Roy/Roving Tortoise Photography, G. R. 'Dick' Roberts/Natural Sciences Image Library. Spine and back cover Tui De Roy/ Roving Tortoise Photography

BLOOMSBURY WILDLIFE
Bloomsbury Publishing Plc
50 Bedford Square, London, WC1B 3DP, UK

BLOOMSBURY, BLOOMSBURY WILDLIFE and the Diana logo are trademarks of Bloomsbury Publishing Plc

First published in 2010 in New Zealand by David Bateman Ltd.
This edition published 2010
Reprinted in 2014, 2016, 2018

A catalogue record for this book is available from the British Library

ISBN: 978-1-4729-6100-6

4 6 8 10 9 7 5

Printed and bound in China through Colorcraft Limited, Hong Kong

MIX
Paper from
responsible sources
FSC® C016074

To find out more about our authors and books visit www.bloomsbury.com and sign up for our newsletters

CONTENTS

Acknowledgements

In a book of this nature, the first acknowledgement must be to the amazing wildlife of Aotearoa New Zealand; thank you for being so patient and understanding.

Starting this book with a very modest knowledge of New Zealand, I have inevitably asked for and received help from a large number of people and institutions: so thanks, first, to the Department of Conservation *Te Papa Atawhai* for all the help, seen and unseen, and for making so many areas accessible and enjoyable; and to the Royal Forest & Bird Protection Society for being such a great champion and supporter of the natural environment.

Over the last four and a half years, I have sought help, information and guidance from many people, among them Jessica Beever, Trent Bell, Hugh Best, John Braggins, Thomas Buckley, Trish Fleming, Graeme Jane, Larry Jensen, Dave Kelley, Brian Lloyd, Peter Newsome, Kelly Ross, Dave Row, Margaret Stanley, Rupert Sutherland, Maggie Wassilief, Janet Wilmshurst and Hugh Wilson. John Sawyer, who helps run the New Zealand Plant Conservation Network was very helpful with plant identification, as was the NZPCN website. My apologies to anyone whom I have missed but who has helped along the way, your contributions, though often quite small, have been invaluable.

It is the photos that make the book, and as such I owe a big thank you to Tui De Roy and Mark Jones from Roving Tortoise Photography, Steve Wood and Peter Smith from NSIL (www.nsil.co.nz), for allowing the use of their photographs at very reasonable rates. My thanks to Tui also for introducing me to New Zealand in the first place.

Finally a very big thank you to Tracey Borgfeldt at David Bateman Ltd., as the deadlines disappeared into the mists of time, she never gave up, patience personified. She has put a huge amount of time and effort into making this book what it is, the first field guide to cover the fauna and flora of New Zealand within a single volume.

Okarito Brown Kiwi *Apteryx rowi*, also known as the Rowi.

Author's Introduction

One of the best kept secrets of Aotearoa New Zealand is its amazing biodiversity, its immense richness of animals and plants and its wonderfully varied geology and geography. When I first arrived here nearly five years ago, I knew very little about the wildlife; I would have been hard pressed to name even half a dozen endemic species, Kiwi, kiwi and er… kiwi! And yet New Zealand possesses a most amazing and unusual fauna and flora.

In talking about the uniqueness of New Zealand, I do not really know where to start. A large, isolated oceanic archipelago, on its own for 70–80 million years, bathed by a cool ocean but blessed by a warm sun. Immensely rich in natural resources, with a hugely varied and unusual geology, snow-capped peaks, rumbling volcanoes, breathtakingly clear air, awesome glaciers and intricately braided rivers. Then there is the wildlife; over 80% of it is endemic, found only here. And not just endemic, but remarkable, very different from anything elsewhere in the world. Sadly, much of it is under threat.

New Zealand has often been referred to as the 'Land of Birds' due to the predominance of birds here, but it is also a land of ferns, of lichens, a land of bugs, a land of trees, a land of reptiles.

So this book is, in a way, my tribute to the native fauna and flora. It is my tribute to those species that have managed to survive the ferocious onslaught of human settlers, both Maori and Pakeha, who throughout their near 750 years on these islands have sought to use and sadly destroy so much of the unique biodiversity that is found here.

Even now, when the problems, dangers and costs of overdevelopment and misuse of natural resources are well known, we continue to rape and pillage the natural ecosystem, safe in the misguided belief that there will always be more, that it will recover, that we will survive, that we know better than nature.

Charles Darwin visited the Bay of Islands in 1835 on his way back to England after four years on a very small, cramped sailing vessel. He was homesick and tired and he dismissed New Zealand, both the wildlife and the people, with some rather uncomplimentary prose. Had he known what we now know about the native wildlife, he would have been mortified and astounded. It was Darwin who first brought the attention of the world to the amazing wildlife of the Galapagos Islands, but what we have here is equally, if not more, amazing, and yet we do not really recognize it for what it is or look after it as the Ecuadorian people look after the Galapagos.

So this book says to everyone, young and old, visitor or resident, Maori, Pakeha, Islander, Asian: This is what you have, did you know that you had it? Go out there and look for it, it is not hard to find, and when you have found it, look after it. It can only survive so much abuse. Do you want to be one of the people who decided that your children or grandchildren should not be entitled to see and enjoy what we can see today?

The natural history, the wildlife, the flowers and trees, the bugs and creepy crawlies, they are all native New Zealanders, they are all part of the rich and diverse culture that is New Zealand. Sing their praises, look after them, appreciate them, they may not look like dollar bills, but if we lose them, then no amount of money will bring them back.

Julian Fitter

MAP OF NEW ZEALAND

North Island

Cape Reinga
North Cape

Ninety Mile Beach

Kaitaia
Kerikeri
Russell
Bay of Islands

Hokianga Harbour

NORTHLAND
Whangarei

Hen and Chickens
Mokohinau Is.

Kai Iwi Lakes
Dargaville

Little Barrier I.
Great Barrier I.

Warkworth
Kaipara Harbour
Kawau I.
Tiritiri Matangi I.
Hauraki Gulf
Cuvier I.
Mercury Is.

AUCKLAND
Auckland

Coromandel

Manukau Harbour

Firth of Thames
Alderman Is.
Slipper I.

Waikato River

Mayor I.

Waihou R

WAIKATO
Hamilton
Tauranga
Bay of Plenty
White I.
Cape Runaway
Motiti I.

Kawhia Harbour
Waikato River
Lake Rotorua
Rotorua
Whale I.
Whakatane

BAY OF PLENTY

North Taranaki Bight

Motu River
Waioeka

EAST COAST

Lake Taupo
Taupo
Rangitaiki River
Lake Waikaremoana
Wairoa
Gisborne

New Plymouth

Mohaka R

Mt Tongariro
Mt Ngauruhoe
Mt Ruapehu
Mt Taranaki

Whanganui River

Hawke Bay
Mahia Peninsula

HAWKE'S BAY
Napier

Rangitikei River
Ngaruroro R

Doubtful Sound

TARANAKI

Hastings
Cape Kidnappers

Whanganui
South Taranaki Bight

MANAWATU/ WHANGANUI

Tukituki River

Dusky Sound

Rangitaiki R
Manawatu River
Palmerston North

Chalky Inlet

Kapiti I.

Cook Strait

WELLINGTON

Ruamahanga R

WAIRARAPA

Lake Wairarapa
Wellington

SOUTH ISLAND

Cook

Cape Palliser

South Island

MAP OF NEW ZEALAND AND ITS OFFSHORE ISLANDS

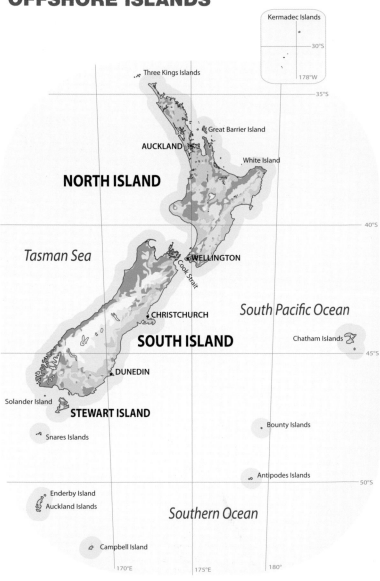

Kermadec Islands

30°S

178°W

Three Kings Islands

35°S

Great Barrier Island

AUCKLAND

White Island

NORTH ISLAND

40°S

Tasman Sea

WELLINGTON

Cook Strait

South Pacific Ocean

CHRISTCHURCH

SOUTH ISLAND

Chatham Islands

45°S

DUNEDIN

Solander Island

STEWART ISLAND

Bounty Islands

Snares Islands

Antipodes Islands

50°S

Enderby Island

Auckland Islands

Southern Ocean

Campbell Island

170°E   175°E   180°

Arawhata River, south Westland.

INTRODUCTION

Overview of natural history

New Zealand has a most unusual biology and geology. It is composed of two very large, and one small, oceanic islands plus a number of offshore islands. Its nearest neighbour is Australia some 2400 km to the west. Because of their isolation the arrival of new species was a rare occurrence and this has resulted in a high incidence of endemism and also of flightlessness in the bird species. With the arrival of Maori some 800 years ago, and, more recently, immigrants from Europe and elsewhere, huge numbers of new species have been brought to New Zealand, both deliberately and accidentally, thus radically altering the ecology of many parts of the country.

Endemism

New Zealand is the most isolated large archipelago in the world, and it has been on its own for some 70–80 million years. This helps to explain the remarkable degree of endemism found here, especially in plant species, with some 84% of flowering plants found nowhere else, and the absence of land mammals, apart from two species of bat. This isolation allowed the evolution of a large number of flightless bird species, the majority of which are now extinct due to habitat loss and predation from humans and other alien species.

Biodiversity

The biodiversity of New Zealand is remarkable. While the islands make up some 0.17% of the earth's land area, it has around 1% of all known land animals; nearly six times as many species as you might expect to find. This is even more notable in view of the fact that isolated areas tend to have lower levels of biodiversity than continental ones. These factors have produced a very unusual collection of birds, invertebrates and plants so that visitors, while being confronted with a number of readily identifiable species, will find others which are so different that they do not know where to start.

Geology

The geology of New Zealand is at least as interesting and rather more observable and identifiable than the wildlife. New Zealand straddles two of the great tectonic plates that make up the earth's crust. The South Island is largely on the Pacific Plate while all of the North Island sits on the Australian Plate. The geology of the two main islands is very complex: the South Island is largely composed of sedimentary and metamorphic rocks while the North Island lacks the schists that make up the Southern Alps, but has extensive areas of young volcanic rocks. The geology has a considerable influence on the fauna and flora, particularly the limestone areas on the South Island; this makes it an important element of the ecology of the islands.

Scenery and Landscape

The combination of plate tectonics, varied geology and oceanic climate have given rise to an amazing variety of dramatic and beautiful landscapes: forest-clad mountains in the far south, the glacier-carved peaks of the Southern Alps (that are still being pushed upwards), the wide open grasslands and immensely wide, braided rivers of Canterbury, the sunken valleys that make up the Marlborough Sounds, the dramatic volcanic peaks of Taranaki, Ngauruhoe, Ruapehu and Tongariro, and the huge expanse of Lake Taupo. Further north, there are the forbidding forests of the Ureweras, with the jewel-like Lake Waikaremoana, lush bush and golden beaches in the Bay of Plenty and Coromandel Peninsula, up to the seemingly endless Ninety Mile Beach of Northland.

New Zealand is a landscape photographer's paradise, with the light and clouds to give mood and expression to the landscape.

Conservation – Introduced and Alien Species

New Zealand developed a very unusual biodiversity. In the almost complete absence of land mammals it developed into a land of birds, many of them flightless. When Maori arrived some 800 years ago, they brought with them the Kuri, or Polynesian Dog, and the Kiore, or Polynesian Rat. Between them they managed to drive at least 31 species of bird, including 11 moas, to extinction. They also, probably through the injudicious use of fire, managed to destroy or seriously alter some 20% of the native bush.

With the arrival of the European settlers in the middle of the 19th century, the rate of destruction of the forest increased exponentially, so that by the end of the 20th century only some 25% of the original native bush remained, most of that seriously compromised by over 30 introduced mammals, both grazers and predators – to say nothing of alien plants and invertebrates which number in their thousands. This has resulted in the loss of another 11 species of native bird, a bat, a fish, at least three frogs and four plants, and unknown numbers of reptiles and invertebrates.

Alien species are expensive; they destroy forest, invade farmland, infect livestock, quite apart from the damage they do to native wildlife and to the wider environment, which has knock-on effects. Only relatively recently has the frontier spirit started to wane, and people have realised that the coastal wetlands, native bush, rivers and alpine areas are not enemies to be defeated but friends who are vitally important to the wellbeing of the whole environment and therefore of the people and the economy of the country.

Partly due to having so many problems with invasive species, New Zealanders have developed a world renowned, and much sought-after, expertise in eradicating noxious pest species. Conservation workers have been particularly successful on offshore islands, of which there are many, and are now developing a network of mainland or inland 'islands', which are helping to ensure that much-threatened native species of birds, reptiles and invertebrates are once again thriving instead of struggling to survive.

At the forefront of the conservation effort is DoC, the Department of Conservation, which manages over 30% of the country. But increasingly, support comes from a large number of private organisations, ranging from the largest national conservation organisation, the Royal Forest & Bird Protection Society, better know as Forest & Bird, to smaller local groups looking after specific areas, such as the Miranda Shorebird Trust on the Firth of Thames, the Maurice White Trust which has developed a remarkable reserve on the Banks Peninsula, and the Karori Wildlife Sanctuary in Wellington which has built a pest-proof fence to develop a wildlife sanctuary close to the heart of New Zealand's capital.

A stoat trap being baited with an egg on Maud Island, one of the many offshore sanctuaries for beleaguered native animals and plants.

The sanctuary at Karori, Wellington, is just one of a number of fenced reserves, the largest being Maungatautari in the Waikato with a 47-km long fence around an extinct volcano. Others are at Bushy Park near Wanganui and the Tawharanui Peninsula north of Auckland. Fencing is very expensive, especially for large areas, and so DoC has developed a series of mainland 'islands' where they use intensive pest control to keep the population of alien predators to a minimum, and these have proved very successful.

A number of offshore islands have also been developed as wildlife sanctuaries: Tiritiri Matangi north of Auckland is possibly the best example and well worth a visit, especially overnight. Here you can start to appreciate the wonderful variety of native birdsong as the earliest naturalists to visit the country, Joseph Banks and Daniel Solander, must have heard it. Others include Kapiti Island north of Wellington and Somes Island in Wellington Harbour, as well as Little Barrier Island near Auckland and Moutohora (Whale) and Tuhua (Mayor) islands in the Bay of Plenty.

There is no realistic chance of getting rid of all, or even any, of the serious pest species from the main islands apart from within fenced areas, and so the fight against them is an ongoing one which deserves the wholehearted support of everyone in New Zealand, whether a resident or a visitor. That way we can ensure that no other native species are driven to extinction and that more and more native wildlife will become increasingly accessible to, and enjoyed by, everyone.

Climate

New Zealand lies right in the Roaring Forties, between 35 and 47 degrees South. In the northern hemisphere that would result in a Mediterranean or continental climate; here it results in what is best described as Oceanic Mediterranean. It varies from subtropical in Northland to cool wet temperate on Stewart Island. The main influences are the oceans and the mountains, with a series of west-moving depressions, especially in the south, and the occasional east-moving cyclone in the far north. These, especially the depressions, are then influenced, at times radically, by the long ridge of mountains running from

Fiordland in the southwest right up to East Cape and the Coromandel Peninsula on the North Island and including the central volcanoes.

With the west winds predominating, the west side, especially on the South Island, is much the wetter, Fiordland recording anything up 12 m (40 ft!) of rain per annum, while less than 100 km to the east, parts of central Otago receive less than 400 mm per annum. The climate in central Otago is in fact best described as continental, with very cold, dry winters and hot, dry summers. These extremes typify the microclimates created by the mountains and found throughout the country. In the North Island the differences are not as marked as the mountains form less of a barrier. Rainfall is much more evenly distributed with warm sunny areas, such as Bay of Plenty and Hawke's Bay, which have around 2500 hours of sunshine a year, receiving up to 1.6 m (60 inches) of rainfall a year. Wind is a major factor as well, with higher western areas being much more windy, and Wellington, due in part to its position on Cook Strait, being the windiest city in New Zealand.

This unusual climate has given rise to very unusual vegetation types. Most of the bush or forest is evergreen temperate rainforest with beech forests predominating in cooler and wetter areas and broadleaf/podocarp forest in warmer drier areas. The latter is particularly luxuriant and hosts an almost tropical variety of trees and shrubs.

The mountains create alpine and subalpine climatic conditions which have resulted in a large diversity of plant species, and a rich and varied vertebrate and invertebrate fauna.

GEOLOGY AND GEOGRAPHY

New Zealand has a wide variety of topographic and geological features: the majestic mountain scenery of the Southern Alps and dramatic snow-capped volcanic peaks in the North Island; the glaciers of the Aoraki Mount Cook massif and hot springs, geysers and other geothermal wonders of Taupo and Rotorua; the soaring sea cliffs of Milford Sound and the wild Fiordland coast to the long sandy beaches of the Bay of Plenty, Auckland and Northland. With hundreds of islands and lakes, braided rivers, ash deposits and dramatic and obvious fault lines, New Zealand has been rightly described as 'a geologist's paradise'.

The key to the geography and geology is the county's position astride the boundary of two of the earth's great tectonic plates, the Pacific and the Australian. The movement of these plates has created, and is still shaping, New Zealand. The Australian Plate is being subducted under the Pacific Plate on the west coast of the South Island, while the reverse is happening on the east coast of the North Island. At the same time, the Pacific Plate is moving north, relative to the Australian Plate, so that over the last 23 million years, the Dun Mountain rocks in northwest Nelson have moved some 460 km from their related rocks in Otago. This plate movement has resulted in the building of the Southern Alps and the development of the Taupo–Rotorua Volcanic area in the north.

The three main islands that make up New Zealand show an amazing diversity of geology. The landscape is young and is constantly changing, being uplifted and eroded at the same time. The Southern Alps are being pushed up by 40–120 mm a year, while the two plates are moving past each other at a rate of 35–50 mm a year. As a result of this very active geology, New Zealand experiences many earthquakes and volcanic eruptions.

New Zealand originally formed part of the supercontinent Gondwana, but broke away from it some 80 million years ago (mya). Since then, the size and shape of New Zealand has varied greatly, from being very largely submerged in the 'Oligocene drowning' some 25 mya, to being an enlarged single land mass during the last ice age, some 20,000 years ago when sea levels were much lower. The oldest rocks date back to the beginning of the Palaeozoic era some 540–570 mya.

There are three basic rock types:

Sedimentary, resulting from the deposition of eroded and other material and which include greywacke, limestone, sandstones, mudstones and conglomerates.

Metamorphic, which are rocks that have been changed by heat or intense pressure, or both, and which include marble, gneiss and various schists.

Igneous, which were originally liquid and include the intrusive plutonic rocks Granite and Diorite and the extrusive volcanic rocks such as basalt, andesite and tuff, this last being actually consolidated volcanic ash.

The geology of the North and South islands is quite distinct. The North Island consists largely of younger volcanics and sedimentaries dating back to the Cretaceous period (65–142 mya), while the South and Stewart islands have a more varied and complex geology, with a core and backbone of very old plutonic and metamorphic rocks surrounded by similar sedimentaries to the North Island, but with outcrops of older sedimentaries (Devonian to Cambrian), pre-Cretaceous volcanics and younger basalts.

The other major topographical and geological feature are the faults, which are related to the edge of the tectonic plates and mark the areas where the plate movement is most noticeable. The main fault, the Alpine Fault, runs in a northeasterly direction just inland from the west coast of the South Island and is clearly visible from the air. It divides in the north Canterbury region and

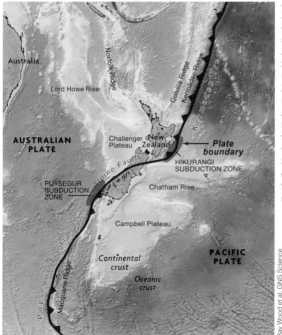

Zealandia, the New Zealand continent, covers 3.5 million sq km, while New Zealand itself is only 268,000 sq km. In the Puysegur Subduction Zone, off Fiordland in the south, the Australian Plate is being forced under the Pacific Plate, while in the Hikurangi Subduction Zone, off the east coast of the North Island, the reverse is the case.

Ray Wood et al, GNS Science

these lesser, but still potentially damaging, faults run up through the central and eastern part of the North Island. The Wairarapa Fault in southern North Island was responsible for the 1855 earthquake that saw the coastline around Wellington rise by up to 6 m. The devastating Napier earthquake of 1931 was caused by another fault, with the land rising by as much as 2.7 m. As well as flattening the city, it drained a lagoon and resulted in 200 ha of new land. In 2010 and 2011, previously unknown faults caused extensive damage in and around Christchurch.

New Zealand has a variety of minerals. Gold was first noted by European settlers in the 1850s, but it was not until 1861 that significant discoveries were made, after which there was a series of 'gold rushes' which were a significant factor in bringing settlers to New Zealand and in opening up the west coast of the South Island. Coal is also found on both North and South islands, and oil and gas are produced in Taranaki with significant possibilities in various offshore locations.

Fossils are another feature of New Zealand's geology and are widespread in the various sedimentary rocks. As recently as 2000, dinosaur fossils were discovered and the petrified forest at Curio Bay south of Dunedin is spectacular and fascinating.

Petrified forest, Curio Bay.

Petrified tree trunk with growth rings visible, Curio Bay.

Sub-fossil molluscs, Whanganui River Gorge.

Boiling mud pool, White Island.

Rocks or 'bombs' eroding out of a pumice and ash cliff, Okurei Point, Bay of Plenty.

Mt Ruapehu as seen from Te Mari, looking across the Blue Lake with Mt Ngauruhoe in between, Tongariro National Park.

Steaming lake, Waimangu Volcanic Valley.

Sulphur and iron deposits, White Island.

Sulphur fumaroles, White Island.

Red Crater, Tongariro National Park.

Moeraki Boulders, coastal Otago.

Inside a Moeraki boulder.

Pancake Rocks, Punakaiki.

Columnar basalt.

Erosion of coastal strata.

Chenier beach, Miranda, Firth of Thames.

Erosion revealing sedimentary rock, Rangitikei River.

Mudstone deposit, Taranaki.

Karst limestone landscape, Nelson.

Glaciers in the Southern Alps, showing lateral and median moraines.

Snowfield at the top of the Fox Glacier.

Fox Glacier.

'U'-shaped valley caused by glacial action.

Braided riverbed, Fox River, Westland.

HABITATS

New Zealand has a very varied topography and a wide range of climates and microclimates. This results in a very large range of habitats, which can, however, be broken down into five main categories.

- Coast and Coastal Wetlands
- Forest – Coastal, Lowland and Montane
- Alpine and Subalpine
- Lakes and Rivers
- Agricultural and Urban

There is considerable variation within these broad categories depending on latitude and rainfall.

Much of the original habitat of New Zealand, especially forest and wetlands, has been destroyed, however, where you find unspoilt or regenerating areas, you will also find the native wildlife, which is, unsurprisingly, better adapted to the native habitat than the introduced species.

Coast and Coastal Wetlands

With over 11,000 km of ocean coastline and 6000 km of sheltered harbour and estuarine coastline, New Zealand has a wide variety of coastal habitats. Surprisingly for a mountainous country, apart from Fiordland in the southwest, there are few sections of coast with high rocky cliffs. Over one-third of the coastline is made up of either sand or shingle beaches. Much of the rest is low tumbling cliffs of a variety of rock types, including long stretches of mudstone or other soft sedimentary material, often eroding rapidly. A good example of this is at Moeraki on the east coast of the South Island where the large spherical Moeraki Boulders are emerging one by one from the mudstone cliffs.

The coast is probably the best place to observe birds as over three-quarters of New Zealand's native breeding birds are found around the coast or on offshore islands and, in addition, coastal habitats are home to tens of thousands of migrant waders, which arrive here each spring from their breeding grounds in the Arctic. Significant numbers of these may overwinter. An added advantage for the visitor, is that the coast and coastal forest is also rather more accessible than inland bush and high alpine regions.

Many sandy beaches are backed by extensive sand dunes which have their own distinctive fauna and flora; species that are perfectly adapted to this quite harsh environment of nutrient poor soils and exposure to wind and salt. The typical plants here are the sand-anchoring grasses such as the natives Pingao *Desmoschoenus spiralis* and Spinifex *Spinifex sericereus,* and the introduced Marram Grass *Ammophila arenaria*. The native grasses, however, are less dense and allow a greater variety of other plants and invertebrates to exist. Sandy beaches, especially in the southeast of the South Island, are favoured by the New Zealand or Hooker's Sealion and are also good places to observe the diminutive Little Blue Penguin which breeds all around the coast.

The other very specific habitat are the many extensive estuaries, found mostly in the North Island. These are excellent places for waders, both residents and migrants, as well as gulls, terns, cormorants and ducks. They also have very distinctive and specialised plant and invertebrate communities and are a vital habitat for native fish. Many freshwater species spawn in the estuaries and have a marine phase in their life, while several marine species also spend time in the estuaries. Estuaries are also an important habitat for shellfish, a source of food for people and birds.

Sand dune with native grasses and invasive lupins.

Pohutukawa on cliff top.

Coastal Kanuka forest, Stewart Island.

Coastal forest, Kapiti Island.

Wind-blown sandy beach, northwest Nelson.

Coastal lagoon, Kapiti Island.

Rocky coasts often have low cliffs, sometimes with native bush down to the water's edge. In the North Island the commonest cliff tree is the Pohutukawa *Metrosideros excelsa*, with its slightly grey-green leathery leaves and brilliant crimson flowers which turn the coast red in December and January. The tidepools and rocks on these stretches of coast are filled with an amazing variety of seaweeds, barnacles, anemones and molluscs.

New Zealand has lost 90% of its coastal wetlands to 'land improvement', mainly for agricultural purposes, but what remains are rich areas frequented by Pukekos, ducks, Fernbirds and Paradise Shelduck and characterised by Cabbage trees, Native Flax and, unfortunately, often the introduced alien Pampas Grass.

Forests

Often referred to as 'the bush', forest once covered 80% of New Zealand. However, the arrival of Maori some 800 years ago and then European settlers in the 19th century, changed that and now only about 25% of the original forest cover survives. This is very generally divided into conifer/broadleaf forest, which is more widespread in the warmer and more fertile areas, and is the predominant forest in the North Island, while beech forest predominates in the South Island, in cooler, wetter climates. Around 98% of the trees and shrubs found in New Zealand forests are evergreen.

Conifer/Broadleaf Forest

Mature conifer/broadleaf forest has five layers:
- Very tall emergent trees, mainly podocarps and other conifers 30–60 m tall.
- A canopy of large broadleaf trees up to about 20 m tall.
- An understorey of smaller trees and tree ferns up to 15 m tall.
- Shrubs and small or juvenile trees up to 3 m tall.
- The forest floor with ferns, grasses, mosses, liverworts and herbaceous plants.

In addition, vines and epiphytes abound, the latter especially in wetter areas where trees can be almost completely hidden by these epiphyte 'gardens'.

The nature and composition of this forest varies considerably with altitude and latitude, ranging from the lowland swamp forests characterised by Kahikatea *Dacrycarpus dacridioides* and Pukatea *Laurelia novae-zealandia*, with their buttress roots and pneumataphores or 'breathing roots', to the impressive Kauri forests of northern North Island and the varied montane forest with a wide variety of species which blends into subalpine scrubland.

Beech Forest

In contrast to coniferous/broadleaf forest, beech forests have very little undergrowth. This is due partly to their being found in cooler and wetter climate zones with relatively poor soils, but also to the lack of light as a result of the very dense evergreen canopy and thick ground coating of very slowly decaying beech leaves. One characteristic of the beech forest is the abruptness of the tree line in subalpine areas, where there is a clear line between forest and grassland with no intervening shrubland.

Mixed broadleaf–podocarp forest showing tree ferns and Rewarewa, Tangarakau Gorge.

Mixed broadleaf–podocarp forest, Pureora Forest Park.

At altitude, beech forest terminates abruptly.

South Island beech forest showing lack of undergrowth.

Alpine and Subalpine

These are some of the least spoilt areas in New Zealand, largely because the land has no agricultural or arboricultural value. However, damage has been, and is still being, done to the ecology of this region by introduced mammals such as Chamois and Himalayan Tahr.

The subalpine areas are often shrublands with several species of hardy evergreens including the graceful Grass Trees *Dracophyllum* spp., several species of Hebe and the distinctive Mountain Pine *Dacrydium bidwillii*, the smallest of the native conifers. There are also large areas of native tussock grasslands of *Chionochloa* spp., which are often dotted with fearsome looking 'Spaniards' or speargrasses *Aciphylla* spp. or large white alpine daisies *Celmisia* spp.

The alpine habitat is characterised by low-growing shrubs and flowers, well adapted to the harsh environment with a mean annual temperature of less than 7.5°C. In spite of this, it has an amazing biodiversity with birds such as the inquisitive Kea or Mountain Parrot *Nestor notabilis* and the diminutive Rock Wren *Xenicus gilviventris*, several species of skinks and geckos, dragonflies and grasshoppers, and over 600 species of plant, some of them ground-hugging shrubs such as Creeping Coprosma *Coprosma pumila* or the curious and aptly named Vegetable Sheep *Raoulia* spp. and the oddly named, scree-loving Penwiper Plant *Notothlaspi rosulatum*.

Beech forest and alpine fellfields and herbfields, northwest Nelson.

Subalpine shrublands, Mt Pureora, central North Island.

Subalpine grasslands.

Fox Glacier, with Mt Tasman.

Hooker Valley dominated by Mt Cook.

Tarn and surrounding bog, Fiordland.

Celmisia, Mt Cook National Park.

Rakaia River, Canterbury.

Lakes and Rivers

Lakes are an important wildlife and plant habitat with over 16 lakes of 50 km^2 or more. In the North Island, lakes are largely of volcanic origin and mainly found in the Taupo/Rotorua volcanic area. In the South Island, they are mostly glacial and are found on the eastern side of the Southern Alps. Lakes are an important habitat and breeding ground for water birds, including terns and cormorants, as well as ducks and swans. They are also an important fish habitat, though the introduction of non-native trout and other species has done some harm to native freshwater fish species.

Being a rather narrow country, New Zealand rivers are not long. The Waikato River, which flows north out of Lake Taupo in the middle of the North Island, is the longest at 425 km followed by the Clutha in the South Island at 322 km.

Many of the rivers, especially where they flow out of the mountains, are several kilometres wide. They are known as 'braided rivers' and have a unique and important, but threatened, ecology. The braided rivers have wide, flat beds which are only rarely filled in severe floods. There are generally two or three smaller channels which crisscross the river bed in a rather random way, frequently changing course and providing excellent breeding habitat for at least 26 species of native birds including the endangered Wrybill *Anarhynchus frontalis*, the only bird in the world with a sideways-bent bill, Black-fronted Tern *Sterna albostriata* and Banded Dotterel *Charadrius bicinctus*. Native plants found include lichens, mosses, willowherbs and grasses, all adapted to the ever-changing environment. Sadly, especially in their lower reaches, many of these rivers have been invaded by non-native plants, especially shrubs and trees, such as willow *Salix* spp., which anchor the sediments and result in permanent islands and channels thus destroying habitat for native species. Another serious threat to the rivers is water extraction for industrial agriculture.

Tokaanu with Lake Taupo in the background, central North Island.

Cardrona River with willows and lupins, Otago.

Waihou River, Hauraki Plains.

Awatere River, Gisborne.

Tarawera River, Bay of Plenty.

Caples River, Otago.

Lake Pukaki, with Mt Cook (centre).

Agricultural and Urban Landscapes

While the area covered by forest has shrunk since the arrival of people, the agricultural and urban environments have expanded from nil to 60%, much of it in the last 160 years. The early European settlers were intent on recreating a western European landscape, both rural and urban. Adding to the destruction of the native flora and fauna, they introduced, both deliberately and accidentally, thousands of new species, many of which have had a devastating effect on native species. However, unlike in Britain and some other parts of Europe, the settlers did not leave small areas of woodland or native vegetation, they went for intensive land usage which has resulted in the complete elimination of native species from many areas.

As a result, in much of the country the commonest trees are European poplars, willows, Macrocarpa, the dreaded *Pinus radiata* and a variety of subtropical ornamental trees and shrubs. You are more likely to see European Blackbirds, Starlings and House Sparrows and Mynas from Asia than almost any native bird species. *Pinus radiata* is extensively farmed and is New Zealand's largest timber export, and while being far less attractive and wildlife friendly than the native bush, it is at least a more environmentally sustainable industry than industrial dairy farming.

A few natives have adapted to this manmade 'wilderness', especially fairly recent arrivals such as the Welcome Swallow and Silvereye, whose arrival is perhaps linked to the development of these new habitats. Many cities have a surprisingly large number of native bird species: Wellington has benefited from the presence of the Karori Sanctuary in its midst, and many other towns have healthy populations of Tui, Fantail, Grey Warbler and Silvereye and well as Kereru and Morepork and even Kaka. Some of these are particularly attracted to spring and summer flowering of nectar bearing trees such as Kowhai, Puriri and Pohutukawa.

Some species have actually benefited from the manmade environment, particularly the Australasian Harrier which is a bird of the open country. The Pukeko *Porphyrio porphyrio* and Paradise Shelduck *Tadorna variegata* have both survived in spite of being the target of hunters. The absence of native plants and habitat also reduces the native invertebrate population, so that most of the insects, spiders and arthropods that you see will be introduced.

So much of the country has been altered that it is only when you finally come across genuine native bush that you can appreciate exactly what you are missing. It is so rich and varied. Make sure that you visit and support as many of the island and mainland sanctuaries and reserves as you can. Here work is being done to turn the tide against the alien invaders. Only in this way is there real hope that future generations will be able to appreciate and enjoy the amazing wildlife of New Zealand.

Recently created dairy pasture on land that was previously commercial forest, Rotorua–Taupo.

Lake Rotoroa, Hamilton.

Wellington city and harbour.

Avon River, central Christchurch.

Beef cattle, North Island hill country.

Sheep farming, north Canterbury.

Horticultural land, Gisborne.

Introduced Gorse, Banks Peninsula.

A Tui on Harakeke, New Zealand Flax. The nectar of its flowers is one of the bird's favourite food sources.

BIRDS

Identifying birds in New Zealand is relatively simple compared to identifying flowers or invertebrates. There are fewer than 300 species that are resident or regular visitors, and an appreciable number of these are found in remote areas, high mountains, offshore islands or inaccessible forests. At the same time, the variety of species is remarkable, ranging from the albatrosses and petrels of the southern ocean, to the flightless Kiwi and Weka, migratory waders such as the Bar-tailed Godwit, introduced European species such as Starlings and Goldfinches, and occasionally in the north, Sooty Terns and Frigate Birds, which have a tropical or subtropical range. Overall, however, NZ birds are adapted to a temperate climate; cool maritime in the south, warm temperate verging on subtropical in the north. Given the space limitations of a guide such as this, not all species have been illustrated or included, but those native species that you are most likely to see are here. Some introduced species are also included.

In order to make the guide easier to use, this section groups birds by habitat. This should make it easier to identify individual species. The groups I have used are:

Seabirds – Albatrosses, Petrels and Shearwaters, Penguins, Gannet and Shags, Gulls, Skuas and Terns

Waterside and Wetland Birds – Herons and Egrets, Swans, Geese and Grebes, Ducks, Rails, Crakes and Moorhens

Waders – Native Breeding and Migratory

Land Birds – Native – Kiwis, Hawks and Owls, Parrots, Pigeons, Cuckoos and Kingfishers, Passerines ('perching' birds) medium-sized, small, very small

Land Birds – Introduced – Game Birds, Passerines

Not all introduced species are in the introduced section as some fit better within the other sections. However, they are clearly identified as introduced or alien species.

In general you will see native land birds in or near the forest or areas of native bush, while the introduced species tend to be found more in open areas where the environment has been radically altered by agriculture and other human activities.

All species will be identified as:

E	Endemic to NZ (found only in New Zealand).
EG	Endemic Genus.
EF	Endemic Family.
EO	Endemic Order.
BE	Breeding Endemic. Breeds only in NZ.
N	Native to NZ. Breeds here & elsewhere, found here year-round.
BM	Breeding Migrant. Breeds in NZ but winters to the north.
NBM	Non-breeding Migrant. Regular migratory visitor but doesn't breed here.
RV	Regular Visitor. Non-migratory and non-breeding.
OV	Occasional Visitor.
I	Introduced or alien species.
IP	Introduced pest or invasive species.

NAMES AND NOMENCLATURE

With the advent of DNA analysis, the species status of many species and subspecies is under constant review. In order to ensure consistency, I have used the BirdLife International list as the main reference point. However, changes are being made all the time. Common names are a bigger problem as the same species may have many different common names in different parts of the world. Here, I have tended to use Maori names for endemic species when they are commonly used (or local English common names) and internationally recognised names for other native and introduced species, with most alternatives listed. Also, all names are indexed, including the scientific names.

CONSERVATION STATUS

For simplicity, I have used the BirdLife International classification system. (The NZ Department of Conservation (DOC) uses a more complicated system to denote conservation status for NZ birds, and other fauna and flora.)

The conservation status of each species is indicated as follows:

CE	Critically Endangered	**T**	Threatened
E	Endangered	**NT**	Near Threatened
V	Vulnerable	**LC**	Least Concern

SIZE: BL = Body length, tip of bill to tail. WS = Wing span.

SEABIRDS

ALBATROSSES

Of the 22 species of Albatross accepted by BirdLife International, 11 breed in New Zealand and its offshore islands and eight are endemic. The 11 species are divided into groups: the three Great Albatrosses, consisting of the Wanderers and Royals; the Mollymawks, which are medium-sized albatrosses, and the Light-mantled Albatross.

THE GREAT ALBATROSSES

These huge birds all have very similar lifestyles and behaviour. All are long-lived (50 years or more), monogamous biennial breeders, though a significant proportion breed less frequently. All have an elaborate and vocal courtship ritual, lay a single white egg and share the incubation and feeding of the chick. Their diets are similar, consisting largely of squid and other crustaceans, with some fish and carrion. They are natural scavengers and at risk from commercial fisheries operations. Mitigation measures in New Zealand waters have helped to reduce the death toll significantly. Courtship and nesting habits are similar; the nest is a low mound of vegetation and some mud, generally in coarse grass or tussock.

Antipodean Albatross *Diomedea antipodensis* **V BE**
WANDERING ALBATROSS, NEW ZEALAND ALBATROSS
TAXONIMIC NOTE: PREVIOUSLY *D. EXULANS* BEFORE IT WAS SPLIT INTO FOUR SPECIES
Best viewed at Kaikoura, on the east coast of the SI.
A huge black/brown and white seabird (BL 110–115 cm; WS 3–3.25 m) with a massive pink bill. Plumage varies with age and sex. M: body white, brown vermiculations on mantle and rump, tail dark sometimes dark cap, upperwing near black, white flecking near mantle. F: much browner, white belly and lower breast. Subspecies *gibsoni* both sexes whiter. Immatures: largely dark brown with white face, throat and underwing, whiter with age. Only 'Wanderer' in NZ, breeds on Antipodes Island (*D. antipodensis*) and in the Auckland Islands (*D.a. gibsoni*) late December through early February. Young fledge Nov–Dec.

Northern Royal Albatross *Diomedea sanfordi* **E BE**
TOROA
Best viewed at Kaikoura, on the east coast of the SI.
A huge largely white seabird (BL 115 cm; WS 3–3.25 m) with black upperwings and dark tips and trailing edge to underwing. Female may have dark cap during breeding season. Bill massive, pink with dark cutting edge to both mandibles. Immature similar but upperwing browner with significant white flecking near mantle, and dark flecking on head, mantle and rump. Breeds on the Forty Fours and The Sisters in Chatham Islands, plus some 30 pairs on Taiaroa Head, Otago Peninsula. Nests on rocky ground and amongst low vegetation. Lays in late October through November. Incubation 60–65 days, fledging Sept–Oct.

Southern Royal Albatross *Diomedea epomophora* **V BE**
TOROA
Best viewed at Taiaroa Head, Otago Peninsula and Kaikoura, on the east coast of the SI.
Slightly larger than Northern Royal (BL 107–122 cm; WS 3.05–3.60 m.) Body white, upperwing dark but white on leading edge of innerwing, broader near mantle. Bill massive, pink, black cutting edge to both mandibles. Female generally less white. Immatures have darker upperwing, white either side of mantle, similar to adult Northern Royal. Breeds at Campbell Island, small colonies in Auckland Islands, and some hybrid northern/southern birds on Taiaroa Head. Lays late Nov–Dec. Incubation 79 days, fledging Oct–Nov.

MOLLYMAWKS AND LIGHT-MANTLED ALBATROSS

The mollymawks are small and medium-sized, largely black and white albatrosses found exclusively in the Southern Hemisphere. There are 10 species, six breed in NZ and five are endemic. They are long-lived (40 to 50 years), monogamous, annual breeders, all breed on isolated islands, build a low truncated nest of mud and vegetation which is re-occupied each year, lay a single white egg, and have a distinctive and at times noisy courtship ritual. They all have similar diets, mainly squid, other cephalopods, crustacea and fish, but as scavengers they are also at risk from commercial fisheries operations. The plumage of all species is very similar, white body and rump, black tail, mantle, upperwing and underwing tips and leading and trailing edges. In most cases the head plumage and bill colouration

Antipodean Albatross
(Adult male)

Antipodean Albatross
(Adult female)

Antipodean Albatross
(Immature)

Northern Royal Albatross

Southern Royal Albatross

Northern Royal Albatross

Northern Royal Albatross

are the distinguishing features. Sexes are similar and immatures generally lack diagnostic features, especially the bill colouration.

White-capped Mollymawk *Thalassarche steadi* **NT** **BE**
WHITE-CAPPED ALBATROSS

Mantle greyer than most mollymawks and distinctive black thumbprint on leading edge of underwing next to body (BL 90 cm; WS 2.12–2.56 m). Head and neck white with grey cheeks, eye dark, dark eye shadow and a distinctive white cap to the head. Bill, pale grey sides, dull yellow ridge and bright yellow tip, bright orange patch at base of lower mandible extends into gape patch visible only when displaying or feeding young. Breeds in Auckland Islands, and on Bollons Island off Antipodes Island.

Salvin's Mollymawk *Thalassarche salvini* **V** **BE**
SALVIN'S ALBATROSS

Light grey head and neck, paler forehead and typical dark triangular eye patch (BL 90–100 cm; WS 2.10–2.50 m.). The bill is dull grey with a pale yellow ridge, horn-coloured tip and a black spot on end of lower mandible. Distinctive black thumbprint on leading edge of underwing next to body. Breeds on Bounty Islands with small population in Snares Group.

Chatham Mollymawk *Thalassarche eremita* **V** **BE**
CHATHAM ALBATROSS

Typical black and white mollymawk (BL 90–100 cm; WS 2.10–2.56 m). Upperwing black, underwing white apart from narrow dark leading and trailing edges. Mid-grey head with dark triangular eye patch and striking custard-yellow bill with dark mark on tip of lower mandible. Breeds on The Pyramid in Chatham Islands. Possibly the smallest breeding area of any bird species.

Buller's Mollymawk *Thalassarche bulleri* **V** **BE**
BULLER'S ALBATROSS

Body and wings as with other mollymawks (BL 78–81 cm; WS 2.05–2.13 m). Head light-grey with paler forehead and crown. Dark eye and triangular eye shadow. Bill black sides, bright yellow upper and lower ridges, orange patch at base of lower mandible extends to gape-cheek patch when feeding and displaying. Two subspecies, Southern *T.b. bulleri* and Northern *T.b. platei*, the latter lacks pale forehead and crown. Breeds on Southern Snares and Solander islands.

Campbell Mollymawk *Thalassarche impavida* **V** **BE**
CAMPBELL ALBATROSS

Typical mollymawk with white body and head (BL 80–95 cm; WS 2.10–2.15 m). Black upperwing, underwing white with wide dark leading and trailing edges. Tail dark. Distinctive bright orange bill and eye golden with dark eye shadow. Immature has darker underwing, grey nape and collar, bill dull with dark tip. Breeds on Campbell Island.

    Similar species: The **Black-browed Albatross** *T. melanophyrs* is similar but with darker eyebrows, dark eye and reddish tip to bill. Small numbers breed on Antipodes, Campbell and Snares islands.

Grey-headed Mollymawk *Thalassarche chrysostoma* **V** **N**
GREY-HEADED ALBATROSS

Black and white mollymawk, (BL 70–85 cm; WS 1.80–2.05 m) head uniform mid-grey with dark eye shadow, bill dark with bright yellow upper and lower ridges. Upperwing dark, underwing white with broad dark leading edge and narrow dark trailing edge. Immature has dull bill with dark tip and lacks white on underwing. Breeds on Campbell Island; the only biennial breeding Mollymawk. Also breeds on South Atlantic and South Indian Ocean islands.

Light-mantled Albatross *Phoebetria palpebrata* **NT** **N**
LIGHT-MANTLED SOOTY ALBATROSS

An elegant all-grey albatross (BL 78–90 cm; WS 1.80-2.20 m) with a noticeably long, pointed tail. Head near-black but neck, mantle and most of body mid to pale grey. Upperwing dark grey, lighter near body, underwing mid-grey. White crescent behind eye. Bill black with blue or violet stripe along lower mandible. Distinctive, haunting, 'peeooo' courtship call. Breeds on Auckland, Campbell and Antipodes islands.

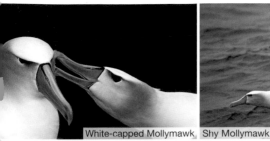
White-capped Mollymawk

Shy Mollymawk

Salvin's Mollymawk

Chatham Mollymawk

Grey-headed Mollymawk

Campbell Mollymawk

Buller's Mollymawk

Light-mantled Albatross

Black-browed Mollymawk

PETRELS AND SHEARWATERS
DARK-PLUMAGED PETRELS
Some 35 species of petrels and shearwaters breed in New Zealand and the offshore islands, and many of these species can be seen when crossing Cook and Foveaux straits or on whale, dolphin or bird-watching trips from Kaikoura and other locations, especially on the east coast. The further out you go, the better your chances. During the breeding season, which is winter for some species, birds congregate offshore at dusk.

Northern Giant Petrel *Macronectes halli* NT N
GIANT FULMAR, NELLY, STINKER
A very large (BL 90 cm; WS 1.50–2.10 cm.) dark grey-brown, albatross-sized petrel with a huge, bulbous horn-coloured bill (9–10.5 cm long), tip darker reddish, very prominent single nostril on the upper ridge. Plumage variable, face and underside quite pale, almost white, eye light. Wingtip more rounded than Mollymawk. Juvenile all dark, near black. Gregarious and aggressive, often feeds on fish offal. Breeds on Auckland, Campbell, Antipodes and Chathams islands.

Similar species: **Southern Giant Petrel** *Macronectes giganteus* is very similar to, but slightly larger than, the Northern Giant Petrel. Paler, especially around head, neck and breast. All-white morph very distinctive. Generally found further south, but regularly recorded in NZ coastal waters, especially on the east coast of both islands. Does not breed in NZ.

Sooty Shearwater *Puffinus griseus* LC N
TITI, MUTTONBIRD
Large, all dark, (BL 44 cm; WS 94–105 cm) sooty-brown shearwater with long narrow wings. Light-grey/silvery area on underwing generally diagnostic. Feet extend slightly beyond tail. Bill slender, dark grey. Breeds in large numbers burrows on Three Kings, Chathams, Stewart, Auckland, Campbell and Antipodes Islands, possibly SI. Some human predation, harvesting of chicks, still practiced.

Wedge-tailed Shearwater *Puffinus pacificus* LC N
MOURNINGBIRD
Medium-sized broad-winged shearwater (BL 46–47 cm; WS 97–99 cm) with long wedge-shaped tail and slightly languid flight. Two morphs, commoner all dark brown with faint 'M' outline on upperwings and mantle. Underwing slightly paler on primaries and secondaries but less so that Sooty and non-reflective. Pale morph mid-brown above, largely white underneath with dark leading and trailing edges to wings. Bill slender, grey. Uncommon around main islands. Breeds on Kermadecs, and throughout Pacific and Indian Oceans.

Flesh-footed Shearwater *Puffinus carneipes* LC N
PALE-FOOTED SHEARWATER
Large all dark shearwater (BL 46–48 cm; WS 1.10–1.20 m) with broad wings and distinctive flesh-coloured feet. Pale rather thin bill with dark tip. Tail rounded. Bill smaller than Westland or Parkinson's. Breeds in burrows on Hen & Chicken, Mercury, Karewa, Saddleback, Trio and Titi islands.

Westland Petrel *Procellaria westlandica* V BE
WESTLAND BLACK PETREL
A large (BL 50–55 cm; WS 1.35–1.40 cm) all sooty black-brown petrel, bill pale horn-coloured with dark tip, upper ridge, and sulcus. Underwing has pale silvery wash in primaries and coverts. Winter breeder in burrows in Paparoa Range, west coast of SI.

Northern Giant Petrel

Northern Giant Petrel

Sooty Shearwater

Wedge-tailed Shearwater

Flesh-footed Shearwater   Westland Petrel

SEABIRDS   **35**

Parkinson's Petrel *Procellaria parkinsoni* **V** **BE**

TAIKO, BLACK PETREL, BLACK FULMAR

Medium-sized all-dark petrel (BL 46 cm; WS 1.15 m), bill pale horn-coloured with black tip, upper ridge and sulcus. Underwing has pale silvery wash on primaries and coverts. Smaller than Westland. Summer breeder in burrows in bush on Little and Great Barrier islands in Hauraki Gulf.

White-chinned Petrel *Procellaria aequinoctialis* **V** **N**

SHOEMAKER, CAPE HEN

Large (BL 51–58 cm; WS 1.34–1.47 m) all dark, sooty black-brown with white patch on chin at base of lower mandible. Bill pale horn-coloured with light tip, dark upper ridge and sulcus. Breeds on Disappointment (Auckland Islands), Campbell and Antipodes islands in burrows often with shallow pool or puddle in front.

Grey-faced Petrel *Pterodroma gouldi* **LC** **BE**

OI, GREAT-WINGED PETREL, *P. MACROPTERA GOULDI*

TAXONOMIC NOTE: PREVIOUSLY CON-SPECIFIC WITH GREAT-WINGED, *P. MACROPTERA*

A medium-sized (BL 40–43 cm; WS 1.02 m) all dark grey-brown petrel with slender body and long narrow wings. Underwing primaries and secondaries slightly lighter. Bill dark but face around bill and chin distinctly lighter. Bill stout, black. Breeds in winter in burrows on cliffs and islands around NI from New Plymouth to Gisborne.

MEDIUM-SIZED BLACK, GREY AND WHITE PETRELS

Grey Petrel *Procellaria cinerea* **NT** **N**

PEDIUNKER, BROWN PETREL, CAPE DOVE

Medium-sized (BL 50 cm; WS 1.15–1.30 m) grey and white shearwater with slightly chunky body. Head and face dark, back and mantle mid-grey, wings long and narrow and somewhat darker, tail wedge-shaped and distinctly darker. Underside of body white, underwing and undertail dark. Bill horn-coloured. Bill stubby, grayish-horn, yellowish tip. Winter breeder in burrows on Campbell and Antipodes islands.

Buller's Shearwater *Puffinus bulleri* **V** **BE**

NEW ZEALAND SHEARWATER

A medium-sized shearwater (BL 43–46 cm; WS 0.96–1.02 m) with distinctive dark 'W' shape on upperwings and mantle. Head and tail dark grey-brown, mantle mid-grey, undersides all white apart from trailing edge of wing and tail. Bill long, slender, pale grey. Breeds in burrows on Poor Knights Islands.

Cape Pigeon *Daption capense* **LC** **N**

CAPE PETREL, PIED, SPOTTED OR PINTADO PETREL

An instantly recognizable black and white rather stubby petrel which frequently follows ships (BL 35–42 cm; WS 80–91 cm). White body with black head and nape, black and white mantle and rump, tail black, upperwing distinctive irregular white patches, underwing white with black leading and trailing edges, undertail mottled, tail black. Bill short, dark. Two subspecies: *D.c. capense* and the breeding endemic *D.c. australe*, or **Snares Cape Petrel**, that breeds on Chatham, Snares, Auckland, Campbell, Antipodes and Bounty islands.

Mottled Petrel *Pterodroma inexpectata* **NT** **BE**

KORURE, RAINBIRD, SCALED OR PEALE'S PETREL

Medium-sized grey and white petrel (BL 34–35 cm; WS 85 cm) with distinctive darker 'M' marking on upperwing. Head, neck, mantle and tail mid-grey, white throat and undertail, grey body, dark eye shadow. Underwings white with distinctive black band running from carpal joint towards, but not reaching, body. Bill short, dark. Breeds in burrows on islands around south of NZ Dec–May, otherwise generally pelagic out of breeding season.

Parkinson's Petrel   Mottled Petrel   Buller's Shearwater

Grey-faced Petrel   Grey Petrel

Cape Pigeon   White-chinned Petrel

Cook's Petrel *Pterodroma cookii* **T E**
TITI

Medium to small petrel (BL 28–30 cm; WS 65–66 cm), from a group known as gadfly petrels. Head and upperparts light grey with dark 'M' marking across back and upperwing. Underside white, except for short black diagonal line across underwing. Face and forehead white with dark patch behind eye. Bill long and fine. Breeds in burrows on offshore islands: Little Barrier and Great Barrier in the north and Codfish Island off the southern coast of SI.

Similar species: **Pycroft's Petrel** *Pterodroma pycrofti* is very similar to Cook's but slightly smaller and darker; upperparts medium grey with 'M' marking across back and wings. Bill short and fine. Breeds in burrows in dense forest on offshore islands of the northeastern coast of NI, including Poor Knights, Hen and Chickens and Red Mercury islands.

Black-winged Petrel *Pterodroma nigripennis* **LC N**
Medium to small (BL 28–30 cm; WS 63–71 cm), distinctive white grey and black petrel, very similar to Mottled Petrel but smaller and lacks grey belly. Dark 'M' marking on wings, white underside with black irregular line on underwing. Bill short, dark. Breeds in burrows or rock crevice on Kermadecs, Three Kings, and Portland and East islands off NI, and Chatham Islands.

Soft-plumaged Petrel *Pterodroma mollis*   **LC N**
SOFT-PLUMAGED FULMAR

Medium to small (BL 32–37 cm; WS 83–95 cm), long-winged petrel. Head, nape, mantle and tail mid-grey, upperwings dark grey. Face, throat and body white with grey collar, sometimes incomplete. Dark eye shadow. Bill short, dark. A dark form exists but is unlikely in NZ waters. The reason for the name is not known as the plumage is no softer than that of other petrels. Breeds in burrows in dense vegetation on Antipodes Island.

Fluttering Shearwater *Puffinus gavia* **LC BE**
PAKAHA, BROWN-BEAKED OR FORSTER'S SHEARWATER

Medium to small short-winged petrel (BL 32–37 cm; WS 76 cm) with a distinctive fluttering flight. Head, nape, mantle rump and tail mid to dark grey, undersides and underwing white with dark tips and trailing edge. Bill slender, dark grey. Often seen close inshore, attracted to fishing boats, strong underwater swimmers using wings. Breeds colonially in burrows on small islands, Three Kings in north to Marlborough Sounds in south.

Hutton's Shearwater *Puffinus huttoni* **E BE**
A small to medium-sized black and white shearwater with short wings (BL 36–38 cm; WS 72–78 cm). Plumage similar to Fluttering but underwing greyer and neck largely grey with only chin white. Bill long and slender, light grey. White on underwing restricted to central area. In flocks during breeding season. Breeds in burrows at altitude of 1200–1800 m in Kaikoura Range, South Island. Seen Kaikoura, east and north coasts of SI and all of NI.

SMALL PETRELS, PRIONS AND STORM PETRELS

Little Shearwater *Puffinus assimilis* **LC N**
ALLIED, GOULD'S OR DUSKY SHEARWATER

Smallest shearwater (BL 27–28 cm; WS 58–67 cm), black and white with short, rounded wings. All black upperparts, all white underside apart from black wingtips and trailing edge. Black on neck extends down to half collar. Legs and feet pale blue. Bill short, slim, dark. Winter breeder in burrows or rocky crevices.
     Several subspecies: *P.a. kermadecensis* in Kermadecs, *P.a. haurakiensis* on Hen and Chicken Islands in the Hauraki Gulf and *P. a. elegans* on Chatham and Antipodes islands.

Fairy Prion *Pachyptila turtur* **LC N**
TITI WAINUI, BLUE OR DOVE PRION, WHALEBIRD

Small, delicate, grey and white petrel (BL 25–26 cm; WS 56 cm). Dove grey above with distinctive black 'M' on wings. Tail black, white eye shadow. Body white, underwing light grey, paler in centre. Bill short and pale grey. Breeds in burrows and rocky crevices on Stewart, Snares and Antipodes islands, and Chatham Islands.

Cook's Petrel   Black-winged Petrel

Soft-plumaged Petrel   Fluttering Shearwater

Fluttering Shearwater   Hutton's Shearwater

Little Shearwater   Fairy Prion

The **Fulmar Prion** *Pachyptila crassirostris* is very similar to the Fairy Prion, though slightly larger (BL 24–28 cm; WS 60 cm) and paler, with a much heavier and broader bill. Tail less black than Fairy Prion. Breeds in crevices and under rocks, on Auckland, Snares, Bounty and Chatham islands.

Broad-billed Prion *Pachyptila vittata* **LC   N**
PARARA, LONG-BILLED PRION. BROAD-BILLED DOVE PETREL, ICEBIRD
Largest of the prions (BL 25–30 cm; WS 57–66 cm) with a black, remarkably duck-like bill. The internal laminae are used to filter plankton. Plumage similar to Fairy and Fulmar prions, but head darker and bill much larger and black. Forehead noticeably steep. Breeds in burrows and rocky crevices on islands off Fiordland, and Stewart, Snares and Auckland islands.

Common Diving Petrel *Pelecanoides urinatrix* **LC   3 subsp: 1 BE, 2 N**
KUAKA, SMALL DIVING PETREL
Unmistakable very small, dumpy, black and white petrel (BL 20–23 cm; WS 33–38 cm). Upperparts all black, body white but with dark extending to form half-collar. Underwing pale grey, bill short and dark. Seems to have trouble getting airborne! Breeds in dense colonies, in burrows and under rocks in tussocky areas. Three subspecies: *P.u. urinatrix* on islands off NI and in Cook Strait; *P.u. chathamensis* on Solanders, Snares and Chathams; *P.u. exsul* on Auckland, Campbell and Antipodes islands.

Grey-backed Storm Petrel *Garrodia nereis* **LC   N**
OCEANITES NEREIS
A small, short-winged storm petrel (BL 16–19 cm; WS 39 cm), dark above but with back grey rather than black. Head and neck black, body white and underwing white with black leading edge and dark tips and trailing edge. Feet extend beyond tail. Bill short and dark. Breeds on Chatham, Auckland and Antipodes islands, forming tunnels in dense tussock vegetation.

Black-bellied Storm Petrel *Fregatta tropica* **LC   N**
Medium-sized (BL 20 cm; WS 45–46 cm), stocky, black and white storm petrel. Head and upperparts black with broad white rump and pale crescentic wing bar. Undersides dark apart from white chin, belly, which has a broken dark line down the centre, and central area of underwing. White on belly connects to white rump. Tail square, feet normally protrude. Bill short, fine and black. Often flies in characteristic zig-zag pattern and in small groups. Breeds on Antipodes, Bounty and Auckland islands and migrates north to tropics in autumn, returning in spring.

White-faced Storm Petrel *Pelagodroma marina* **LC   N**
TAKAHIKARE-MOANA, FRIGATE PETREL, MOTHER CAREY'S CHICKEN
A large long-legged storm petrel (BL 18–21 cm; WS 42–43 cm), generally mid to dark grey above and white below, dark cap with white above eye. Grey rump, black tail. Primaries and secondaries and tail all black. Body and underwing white but broad dark wing tips and trailing edge. Breeds in burrows in colonies often in sandy soil. Bill short, black.Two NZ subspecies: *P.m. maoriana* on islands off both North and South islands, Chatham Islands and Auckland Islands; *P.m. albiclunis* on Kermadec Islands.

New Zealand Storm Petrel *Pealeornis maoriana* **CE   E**
A small to medium-sized, slim-winged storm petrel (BL 17 cm), with long legs and feet, which project well beyond the square, slightly notched, tail. Head, neck and upperparts blackish-brown except for pale carpal bar; rump white. Breast blackish-brown grading into blackish streaks on white belly, flanks and undertail coverts. Underwing dark with pale central patch. Bill, eye, legs and feet black. Flight is swift-like with alternating flapping and glides. Thought extinct, but observed in the Hauraki Gulf since 2003; breeding site unknown.

Broad-billed Prion

Broad-billed Prion

Common Diving Petrel

Grey-backed Storm Petrel

White-faced Storm Petrel

New Zealand Storm Petrel

Black-bellied Storm Petrel

PENGUINS

While 13 species of penguin have been recorded in New Zealand, only six species breed here and are likely to be observed.

Yellow-eyed Penguin *Megadyptes antipodes* **V E**
HOIHO

SI and Stewart Is., especially Oamaru, Taiaroa Head on Otago Peninsula, Caitlins and Milford Sound.

Medium-sized yellow-headed penguin (BL 56–78 cm), dark black-grey above, white underneath. Crown yellow with black streaks (feather shafts). Bright yellow band from eye to eye round back of head. Paler yellow chin and cheeks, bill orange-yellow with red-orange tip. Feet dull pink. Immatures duller and lack yellow head band. Breeds in burrows close to shore, south and southeast of SI, plus Stewart, Auckland and Campbell islands.

Little Blue Penguin *Eudyptula minor* **V N**
KORORA, LITTLE SOUTHERN BLUE (AUSTRALIA), FAIRY PENGUIN

Widespread around coasts. Especially Bay of Islands, Kawau and Mercury islands, Wellington Harbour in NI and Banks Peninsula and Milford Sound in SI, also Stewart Is.

The smallest penguin (BL 40–45 cm), mid blue-grey above, white below, bill dark, eye light. No distinctive makings. Breeds in burrows, often some way inland, all round coast of all three islands and in the Chathams.

Fiordland Crested Penguin *Eudyptes pachyrhynchus* **V E**
TAWAKI, FIORDLAND OR THICK-BILLED PENGUIN

SI coast, Fiordland, Milford Sound, Stewart Island.

Small black and white penguin (BL 55 cm), black above, white below, head black with distinctive yellow eyebrow extending back to become a short plume; lighter cheek patch may appear as white lines. Bill orange, feet pink. Breeds individually in holes and caves, in coastal forest around southwest coast of SI.

Snares Crested Penguin *Eudyptes robustus* **V E**
SNARES PENGUIN

Snares Islands, and in late summer or autumn on Stewart Island and southern SI.

Very similar to Fiordland Crested but smaller (BL 51–61 cm) and with longer drooping eyebrow or crest. Pink skin at base of bill. Breeds in colonies in clearing or among boulders in Snares Islands. Occasional visitor to main islands.

Erect-crested Penguin *Eudyptes sclateri* **E E**
BIG CRESTED OR SCLATER'S PENGUIN

Antipodes Island, south coast of SI in autumn and winter.

Similar to Fiordland but larger (BL 67 cm), with distinctly erect yellow eyebrow or crest and with whitish skin at base of bill. Bill larger and heavier than Fiordland. Breeds in open colonies among rocks on Antipodes and Bounty islands. Regular visitor to main islands in winter.

Eastern Rockhopper Penguin *Eudyptes chrysocome filholi* **V N**
CRESTED, TUFTED OR VICTORIA PENGUIN, ROCKY, JUMPING-JACK

Breeding islands and around southeast coast of the SI.

Similar to Fiordland but with much larger drooping yellow head plumes and pink skin at base of very stout orange bill. Smallest of crested penguins (BL 45–58 cm). Breeds in open colonies in rocks on Antipodes, Snares, Auckland and Campbell islands. Occasional visitor to main islands.

Yellow-eyed Penguin

Little Blue Penguin

Erect-crested Penguin

Fiordland Crested Penguin

Snares Crested Penguin

Eastern Rockhopper Penguin

GANNET AND SHAGS

Australasian Gannet *Morus serrator* LC N
TAKAPU, BOOBY, SOLAN GOOSE
Widespread around coast in spring and summer.
Large white and black seabird (BL 85–94 cm; WS 1.7–2.0 m) with creamy yellow head and large pointed blue-grey bill. Primary and secondary feathers black as well as central tail feathers. Juvenile mottled brown and white, attains adult plumage 3–4 years. A plunge diver. Breeds colonially mainly on offshore islands but also on NI at Muriwai on the west coast of Auckland and Cape Kidnappers on the east coast in Hawke's Bay; and in the SI at Farewell Spit in the north and Nugget Point in Southland. Nest is made of seaweed and guano on bare exposed rock. Nests July–Feb.

Black Shag *Phalacrocorax carbo* LC N
KAWAU, BLACK CORMORANT, GREAT CORMORANT, CARBO
Widespread by water throughout the country.
Large black-brown water bird (BL 80–88 cm; WS 1.3–1.5 m) with long neck, bill long, hook-tipped grey becoming yellow at base. Facial skin yellow with white cheek patches. Breeding adult has small crest, white plumes on neck, a green gloss to plumage and white thigh patches. Dives from surface, body sits very low in water. Breeds in trees or rock ledges, by rivers and lakes, and sheltered coasts. Breeding time varies according to location.

Pied Shag *Phalacrocorax varius* LC N
KARUHIRUHI, PIED CORMORANT
Warmer sheltered coasts, around north of NI from Raglan to Hawke's Bay; in SI, Tasman and Golden bays, Marlborough Sounds, Banks Peninsula, Fiordland, also Stewart Is.
Large black and white shag (BL 65–85 cm; WS 1.1–1.3 m), crown, nape, back, upperwing and tail black, body and face white apart from broad dark line on flanks to legs. Bill long, grey, lower mandible becoming pinkish at base, yellow patch in front of eye. Breeds colonially, year-round in trees along cliffs close to coast.

Little Black Shag *Phalacrocorax sulcirostris* LC N
KAWAUPAKA, LITTLE BLACK CORMORANT
Widespread around coasts and inland wetlands and rivers in NI. Quite common in northern coastal areas in SI.
Small all black shag (BL 55–65 cm; WS 0.95–1.05 m), with green gloss to plumage. Bill mid-grey, long and slender. Facial skin dark, eye green. Gregarious. Breeds in large colonies in trees, occasionally on ground, near harbours, estuaries and inland lakes.

Little Shag *Phalacrocorax melanoleucos* LC N
KAWAUPAKA, WHITE-THROATED SHAG, LITTLE PIED CORMORANT
Widespread on all three islands, around the coasts and on inland wetlands, lakes and rivers.
Smallest shag (BL 55–65 cm; WS 85–90 cm). Black head, nape, back and tail but white varies: i) White face and throat; ii) White face, throat, breast and body; iii) White face, throat and partially white body. Tail long, bill, short, hooked and yellow. Lighter phases more common in north. Breeds in colonies in trees near coast, estuaries, inland lakes and rivers.

Spotted Shag *Phalacrocorax punctatus* LC E
PAREKAREKA, BLUE SHAG, FLIP-FLAP
Mainly SI, especially north and east coasts. In NI found mainly around Wellington and Auckland.
Medium-sized slender shag (BL 64–74 cm; WS 91–99 cm), with distinctive varied plumage. Wings and mantle mid-brown with spots, breast and body slightly pinkish grey. Head and neck near black with grey line running from above eye to breast. Near black tail, rump and thighs. Legs yellow, bill long, slender, yellowish brown. In breeding season grey neck stripe is larger and white, head and neck develop distinct double crest with white plumes on side of neck and thighs. Breeds in small colonies on cliffs and ledges around coast. Year-round breeder.

New Zealand King Shag (*P. carunculatus*) is a large black and white shag with pink feet and conspicuous yellow caruncles in front of eye. It is found only in Marlborough Sounds. The **Stewart Island Shag** (*P. chalconotus*) is similar to the King Shag but has reddish facial

Australasian Gannet

Little Black Shag

Black Shag

Little Shag

Spotted Shag

Pied Shag

skin, it also has all brown 'Bronze Phase'. It is found on Stewart Island and on the southeast coast of the SI. **Campbell Island Shag** (*P. campbelli*): as King Shag minus caruncles; **Auckland Island Shag** (*P. colensoi*) as Campbell; **Bounty Island Shag** (*P. ranfurlyi*) as Campbell; **Chatham Island Shag** (*P. onslowi*) as King Shag but caruncles large and prominent; **Pitt Island Shag** (*P. featherstoni*) similar to Spotted Shag.

GULLS, SKUAS AND TERNS

Arctic Skua *Stercorarius parasiticus* LC M
ARCTIC JAEGER, PARASITIC SKUA OR JAEGER
In coastal waters Nov–May. Often seen pursuing terns and gulls.
Medium-sized, long-winged, long-tailed seabird (BL 46–67 cm; WS 97–115 cm, female larger). Dark morph: all dark brown with black cap, grey nape. Light morph: white throat, nape and abdomen, grey breast band, white barring on rump, white flash at base of primaries on underwing. Breeding plumage more black and white with cream nape and long central tail feathers. Breeds in Arctic North America and Asia.

Similar species: **Pomarine Skua** *S. pomarinus* is larger with broader wings.

Brown Skua *Catharacta lonnbergi* LC N
HAKOAKOA, GREAT, SOUTHERN, SUB-ANTARCTIC SKUA, SEA-HEN
Seen near breeding grounds then disperses to mainland, especially south SI, during the winter.
Large, noisy, aggressive seabird (BL 64 cm; WS 1.47 m). Plumage uniformly dark brown, large powerful bill with hooked tip. In flight shows white flashes on wings at base of primaries. A scavenger, attacks smaller birds, very aggressive when breeding. Previously considered a subspecies of the Northern Hemisphere *C. skua*. Breeds on Chatham, Stewart, Solander and sub-Antarctic islands on ground in rough grass and tussock.

Similar species: **South Polar Skua** *C. maccormicki*. Smaller and rather yellow on nape, white flashes on underwing very prominent. Bill short and black.

Southern Black-backed Gull *Larus dominicanus* LC N
KARORO, NGOIRO (IMM.), KELP GULL (AUSTRALIA), DOMINICAN GULL
Widespread around coasts and inland throughout the country.
Large black and white gull (BL 49–62 cm; WS 1.06–1.42 m). All white apart from upperwing and mantle which are black with white trailing edge. Large yellow bill, tip hooked, red spot on lower mandible near tip. Legs dull yellow. Juvenile all mid-brown, develops adult plumage over 3–4 years. Breeds on ground, in colonies or solitary, on coast, in mountains to 1500 m, or urban areas.

Red-billed Gull *Larus novaehollandiae* LC N
TARAPUNGA, SILVER GULL (AUSTRALIA), RED-LEGGED GULL
Widespread especially out of breeding season, coast, inland and urban.
Small grey and white gull (BL 36–44 cm; WS 91–96 cm), all white apart from upperwing which is grey with black and white tips. Bill and legs bright red. Eye yellow with red ring. Gregarious, well adapted to human activities, very vocal at times. Breeds in densely packed colonies, around coast, also Lake Rotorua.

Black-billed Gull *Larus bulleri* E E
BULLER'S GULL
Widespread, inland in spring and summer in SI, coastal wetlands in autumn, winter. Small, isolated colonies in NI.
Small very pale gull (BL 35–38 cm; WS 81–96 cm); very similar to Red-billed but smaller and upperwing pale grey with black and white tips. Bill black and rather thin, slightly longer than Red-billed. Eye white, ring black. Legs dull red-black. Breeds on ground in braided riverbeds in SI. In NI breeds occasionally on coastal spits.

Arctic Skua

Brown Skua

Juvenile Southern Black-backed Gull

Southern Black-backed Gull

Red-billed Gull

Black-billed Gull

Caspian Tern *Sterna caspia* LC N
TARANUI
Widespread NI and SI. Mainly round coasts and in estuaries. SI birds tend to move north in winter.
Very large tern (BL 47–54 cm; WS 1.32–1.45 m), all white apart from light grey upperwing with dark tips. Face and cap black. Tail short and modestly forked. Bill very large and bright red with dark terminal band and pale tip. Legs black. Winter plumage head loses blackness, bill duller. Breeds on ground in loose colonies on estuarine sandbanks, occasionally inland in riverbeds.

White-fronted Tern *Sterna striata* LC N
TARA, KAHAWAI BIRD, BLACK-BILLED OR SOUTHERN TERN
Widespread around coasts.
Common very pale tern (BL 35–43 cm; WS 79–82 cm). All white apart from upperwing and mantle which are light grey; slightly darker wingtips. Cap and nape black extending over eye. Bill black, long and very pointed; legs black. Non-breeding plumage: black cap recedes and is grey around eye. Breeds on ground, in colonies on coastal beaches, sandspits and rocky promontories.

Similar species: **Antarctic Tern** *S. vittata bethunei* is slightly larger, greyer and in breeding plumage has a black cap and bright red bill. It breeds on Stewart Island and the sub-Antarctic islands but is not recorded breeding on the mainland.

Black-fronted Tern *Sterna albostriata* E E
TARAPIROE
Inland on SI near rivers. In autumn and winter on east coast including east and west coast of NI to Bay of Plenty.
Small, all grey body (BL 30 cm; WS 65–72 cm) apart from undertail and rump. Head jet-black but white below eye. Bill and legs orange-red. Non-breeding plumage: loses most of black cap and bill has dark tip. Breeds on ground in braided riverbeds in small colonies in SI only. Nests Oct–Feb. Young remain close to nest, fledge 4–5 weeks.

Fairy Tern *Sterna nereis* V N
Breeds Northland, mainly east coast. Overwinters on Kaipara Harbour, with some overwintering Firth of Thames and Bay of Plenty.
Very small (BL 25 cm), underparts and neck white, upperwing and mantle pale grey, tail deeply forked, head black with white forehead, bill and legs orange. Non-breeding bill more yellow with dark tip, black on head less distinct. Immature dark leading edge to wing, dark bill and legs.
    NZ subspecies *S.n. davisae* is critically endangered with a population of fewer than 100 birds.

Grey Ternlet *Procelsterna cerulea* LC N
BLUE OR GREY NODDY
Breeds Kermadecs and has bred on islands off north and east of NI.
Small (BL 28 cm), delicate grey tern. Head and body pale bluish-grey, wings darker, especially primaries, head and nape whiter, tail deeply forked. Eye black with dark mark in front and white behind. Bill black and finely pointed, legs and feet grey.

Little Tern *Sterna albifrons* LC RV
Harbours, estuaries and coastal lakes, mainly NI, summer only.
Very small (BL 25 cm), rather noisy tern. Non-breeding plummage: white head and undersides, black, or near-black cap and nape. Upperwing mid-grey, underwing pale grey with darker primaries. Bill finely pointed, dull yellow at base, tip dark. Legs dull brown-yellow; tail deeply forked. Breeding adult heavier black cap and lores; bill bright yellow apart from black on very tip. Voice a noisy 'keek, keek, keek' or more racous 'kreeek'.

Caspian Tern

Black-fronted Tern

White-fronted Tern

Fairy Tern

Grey Ternlet

Little Tern

WATERSIDE AND WETLAND BIRDS
HERONS AND EGRETS

Kotuku *Egretta alba* LC N
WHITE HERON, GREAT EGRET, GREAT WHITE EGRET/HERON
Okarito during breeding season, disperses throughout country, mainly coastal and estuarine wetlands, occasionally inland lakes and lagoons.
Largest native heron (BL 83–103 cm; WS 1.5 m), all white with long, yellow, dagger-shaped bill, legs black, lighter towards body. Breeding plumage adds elegant dorsal plumes or 'aigrettes' and bill turns black. Distinctive kink in neck. Breeds in trees, single colony in Okarito Lagoon on the west coast of the SI.

Cattle Egret *Bubulcus ibis* LC N/RV
BUFF-BACKED HERON
Widespread especially in agricultural areas.
Stocky all white heron (BL 70 cm; WS 88–91 cm), with long, dagger-shaped, yellow bill and grey-black legs. In breeding plumage it develops golden-buff coloration on head, neck and back, and bill and legs develop orange-reddish tinge. Not known to breed in New Zealand but recorded year-round (first recorded 1963).

Little Egret *Egretta garzetta* LC OV
Occasional visitor to coastal lakes and estuaries, NI and SI.
Small (BL 60 cm) elegant, all-white heron with long, slender, sharply pointed black bill. Yellow lores and eye-ring. Legs black but with yellow soles to the feet. Often observed chasing fish with wings partially open.

White-faced Heron *Ardea novaehollandiae* LC N
BLUE CRANE
Widespread in both coastal and inland wetlands. Often flocks in winter.
Commonest heron in NZ, medium-sized (BL 66–68 cm; WS 106 cm) but very slim. Plumage largely light to mid-grey but with buff breast and body and white throat and face. Bill black, legs greenish-yellow. Long plumes on back, more prominent in breeding season. Immature similar but lacks white face. Breeds (first recorded 1941) in small colonies, high up in tall trees, often pines or eucalypts, mainly coastal wetlands.

Reef Heron *Egretta sacra* LC N
MATUKU MOANA, EASTERN HERON (AUSTRALIA) OR BLUE REEF EGRET, BLUE HERON
Widespread around coast and coastal wetlands especially Northland and Auckland, less common further south.
Medium-sized (BL 60–66 cm; WS 90–100 cm) all slate-grey heron. Long, dagger-like, dark horn-coloured bill, legs short (for a heron) and greyish yellow-green. Grows long plumes on back and neck in breeding season. Immature browner. Flight slow and close to the water. A solitary nester along coast in crevices and on rock ledges, holes and tree roots.

Australasian Bittern *Botaurus poiciloptilus* E N
MATUKU, MATUKU-HEREPO, BROWN BITTERN, BLACK-BACKED BITTERN
Widespread, but secretive, by inland lakes and swamps with reed beds especially northern NI.
Medium-sized (BL 66–76 cm; WS 1.05–1.18 m) long-legged all-brown bird. Plumage heavily streaked and patterned for camouflage, lighter on throat, darker on crown, nape and back. Neck shorter than heron. Bill medium-sized, dagger-like, orange-brown. Distinctive 'booming' call in breeding season Sept–Nov. Breeds in reed beds in or by water.

Royal Spoonbill *Platalea regia* LC N
KOTUKU NGUTU-PAPA
Coastal lagoons and wetlands, NI and SI.
Large, all white, long-legged bird (BL 74–81 cm; WS 1.2 m) with unmistakable large, laterally flattened, black bill 18 cm long. Black facial skin with red spot on forehead. Legs black. Breeds in small colonies in tall trees near water, Kapiti Island and coastal lagoons in southeast of SI.

Kotuku   Cattle Egret   White-faced Heron

Reef Heron   Royal Spoonbill

Little Egret   Australasian Bittern

SWANS, GEESE AND GREBES

Black Swan *Cygnus atratus* **LC I/N (Aust)**
Widespread on lakes and ponds, also coastal, NI and SI.
Very large black bird (BL 1.1–1.4 m; WS 1.6–2.0 m) with very long neck. Distinctive bright red bill with white lateral band near tip. In flight displays prominent white primaries and secondaries. Juvenile dark brown. Breeds singly or colonially, large mound of grass, raupo (bulrush) and flax, within 100 m of lakeshore. Introduced, but may also have arrived naturally, a near-pest species in some areas.

Also: The **Mute Swan**, *C. olor*, is larger and all white. It has a pink bill with a prominent black knob at its base. A European introduction, it is widespread but few in number.

GEESE
There are no native geese, but three introduced species. The largely grey **Cape Barren Goose** *Cereopsis novaehollandiae* is rare, while the heavily bodied and largely white **Feral** or **Farmyard Goose** *Anser anser* is widespread and is often found well away from water. The commonest of the three is the **Canada Goose** *Branta canadensis* which is found everywhere apart from the far north and is clearly identifiable by its brown body and wings and black neck and head with a white band behind the eye.

Paradise Shelduck *Tadorna variegata* **LC E**
PUTANGITANGI
Widespread in pastures and wetlands throughout New Zealand.
Large goose-like duck (BL 63 cm). Male finely barred dark brown, head and neck black with metallic green sheen, prominent white patches on wing in flight, chestnut undertail. Bill and feet black. Female largely chestnut with distinctive white head, bill and feet black, white speculum in flight. Normally in pairs, but flock during moult. Breeds in fields near water, occasionally in holes in trees. Female incubates and cares, male stands guard.

Also: **Chestnut-breasted Shelduck** *T. tadornoides*, similar in size but both sexes have dark heads and chestnut neck and breast. Irregular visitor, known to have bred twice.

GREBES

Australasian Crested Grebe *Podiceps cristatus* **LC N**
PUTEKETEKE, KAMANA, GREAT CRESTED GREBE
Mainly subalpine lakes in SI: Alexandrina, Kaniere, Ianthe, Maporika, Paringa, Te Anau, Manapouri and the Eglinton Valley, also Canterbury lakes and Arthur's Pass.
Large (BL 48–60 cm), quite secretive, elegant grebe with distinctive crest. On water appears all mid to dark-brown with white throat, chestnut ruff and white face patch. Crest dark chestnut, bill long and dagger-like with dark lores. In flight prominent white patches fore and aft on inner upperwing. Immature paler with white stripes up neck, lacks crest. Sits low in the water. Has notable courtship 'dance' and makes a floating nest in reeds or willow branches; striped young leave nest on hatching.

New Zealand Dabchick *Poliocephalus rufopectus* **V E**
WEWEIA
Widely distributed on lakes and ponds. mainly in east and north of NI: Lakes Owhareiti and Omapere, Kereta; Matata Lagoon and on lakes around Rotorua. No longer found in SI.
Small, dark brown grebe (BL 29–30 cm). In water, dark back, paler flanks, neck dark on back, rich chestnut on front, throat paler, chin dark with highlights. Head glossy black with green sheen and white highlights. Pale eye, short dark pointed bill. Non-breeding plumage much drabber. White on wings visible in flight only. Breeds on the edge of lakes and lagoons, floating nest attached to vegetation.

Smaller than the Dabchick is the **Australasian Little Grebe** *Tachybaptus novaehollandiae* which is largely in the far north, it has a distinctive yellow mark at the base of the bill and a rufous patch on the neck in breeding plumage. Often seen on lakes Owhareiti, Omapere and Kereta in the north and on St Anne's Lagoon in the south.

Black Swan   Cape Barren Goose

Feral or Farmyard Goose   Canada Goose

Australasian Crested Grebe

New Zealand Dabchick   Paradise Shelduck

DUCKS

Blue Duck *Hymenolaimus malacorhynchos* **E EG**
WHIO (PRONOUNCED FEE-O)
Distribution throughout upland areas in central NI and western SI on fast-flowing streams and rivers.
Medium-sized duck (BL 53 cm) with all grey plumage with bluish tinge, breast has reddish spotting, less obvious in female. Bill pale pinkish white and flexible. Breeds in hollow logs or other cavities near fast-flowing upland streams, very territorial. Young have extra large feet and leave nest soon after hatching.

Grey Duck *Anas superciliosa* **LC N**
PARERA, PACIFIC BLACK DUCK, BLACK DUCK (AUSTRALIA)
Widespread in both islands; small lakes, slow rivers, estuaries.
Medium-sized, all-brown duck (BL 47–60 cm) with distinctive head markings. Dark crown and eye stripe, with cream colouring either side. In flight green speculum visible. Bill grey. Commonly hybridises with mallard. Breeds in dense vegetation or in trees not far from water. Often two broods in a season.

New Zealand Shoveler *Anas rhynchotis variegata* **LC N**
KURUWHENGI, SPOONBILL, SPOONIE, AUSTRALASIAN SHOVELER
Widespread especially on lowland lakes and in wetlands and sewage farms.
Medium-sized colourful duck (BL 45–55 cm) with large, grey, strikingly spatulate bill. Male chestnut flanks with white spot aft. Back variegated when at rest, head and neck slate grey, white crescent in front of eye, which is bright golden-yellow. Breast variegated grey-brown. In flight, outerwing brown, with innerwing grey leading edge, green trailing edge and white line between. Female generally mottled mid-brown all over. Breeds in thick grass or rough pasture. Female incubates while male stands guard.

Grey Teal *Anas gracilis* **LC N**
TETE, SLENDER, WOOD, OCEANIC OR MOUNTAIN TEAL
Widespread throughout NZ especially on lowland wetlands and occasionally in estuaries.
Small duck (BL 42–44 cm) with mottled light brown plumage, darker on crown and back, but much paler on throat and cheeks. In flight, speculum has dark green sheen. Breeds in dense vegetation or hole in a tree close to water.

Brown Teal *Anas chlorotis* **E E**
PATEKE, BROWN DUCK
TAXONOMIC NOTE: *A. AUCKLANDICA* HAS BEEN SPLIT INTO THREE SPECIES: *AUCKLANDICA* ON AUCKLAND ISLANDS; *NESIOTIS* ON CAMPBELL ISLAND; AND *CHLOROTIS* ON NI AND SI.
Once widespread, now reduced to a few locations. NI: Mimiwhangata, Whananaki, Tawharanui, Cape Kidnappers and Great Barrier, Little Barrier and Tiritiri Matangi islands. SI: Possibly in Fiordland.
Small brown duck (BL 48 cm). Male has dark green head with white eye ring, chestnut breast and flanks, dark brown wings and body, green speculum in flight. Female all dark brown, mottled apart from head, white eye ring. Bill slate grey both sexes. Breeds in dense vegetation, fern and tussock, close to water.

New Zealand Scaup *Aythya novaeseelandiae* **LC E**
PAPANGO, BLACK TEAL
Widespread, locally common, on larger lakes on NI & SI.
Small, dumpy all dark diving duck (BL 40 cm). Male near black above with purplish sheen turning greenish on head. Eye golden, bill grey. Dark brown underneath and white primaries and secondaries visible in flight. Female dark brown with off-white marking at base of bill on some individuals. Breeds in dense vegetation close to water. Second brood common.

The commonest duck in New Zealand is the introduced **Mallard** *Anas platyrhynchos*. The male has a shiny dark green head and neck, chestnut breast with white ring and a yellow bill. The female is largely mottled mid-brown. In flight, both sexes show bright blue speculum. It is found almost everywhere except at altitude and readily hybridises with the native Grey Duck.

Blue Duck   Grey Duck

New Zealand Shoveler   Grey Teal

Brown Teal   New Zealand Scaup (female)

Mallard (male at right)   New Zealand Scaup (male)

RAILS, CRAKES AND MOORHENS

Weka *Gallirallus australis* V E
KELP HEN, WOODHEN, WOODRAIL
Mainly east and north of NI, north and west of SI also Stewart Is.
Large, flightless, brown rail (BL 46–60 cm) with strong dagger-like bill. Plumage largely red-brown with dark streaks, bill and legs orange-brown. Pale band above the eye. Four subspecies: **North Island** *G.a. greyi*; **Western** *G.a. australis*; **Buff** *G.a. hectori* and **Stewart Island** *G.a. scotti*; different races vary in plumage and size. Territorial, breeds in secluded and sheltered locations; hollow logs, short burrows, etc. Year-round opportunistic breeders.

Banded Rail *Gallirallus phillippensis* LC N
MOHO PERERU, BUFF-BANDED RAIL, PAINTED RAIL, STRIPED RAIL
Secretive, by lakes, wetlands and mangroves on all three main islands.
Medium-sized rail (BL 30–33 cm), distinctive white barring on breast and abdomen with chestnut patch on breast. Throat barred grey, head dark brown with prominent white stripe above eye. Back mid-brown with black and white flecking. Territorial; breeds in thick grass or rushes near water.

Spotless Crake *Porzana tabuensis* LC N
PUWETO, PUTOTO, SOOTY RAIL, SWAMP CRAKE, MOTORCAR BIRD
Mainly NI and north & southeast of SI, in raupo and sedge swamps.
Small, very dark rail (BL 17–20). Almost all black but brownish on back/upperwing, and dark grey underneath. White barring in undertail. Legs reddish, bill short, sharp and black, red eye ring. Hard to see but listen for calls like a motor car or alarm clock! Very secretive and territorial, and breeds in thick vegetation close to water.

Marsh Crake *Porzana pusilla* LC N
KOTAREKE, BAILLON'S CRAKE, LESSER SPOTTED CRAKE
Widespread in or near wetlands, NI and SI.
Smallest NZ rail (BL 15–18 cm), rust-brown above with black and white mottling and flecking, especially on wings. Underparts mid-grey, flanks and undertail barred black and white. Legs pale greenish-grey, bill dark brown. Hard to observe. Secretive and rarely observed. Nests in rushes near water.

Pukeko *Porphyrio porphyrio* LC N
PURPLE GALLINULE, PURPLE SWAMPHEN, BLUE BALD COOT
Widespread, especially open pastures by ponds and streams.
Large, very long-legged, all dark bird (BL 44–48 cm) with dramatic scarlet bill and frontal shield. Head and upperparts black with greenish sheen; neck, breast and abdomen bluish-purple, thighs black, undertail white; legs red. Breeds in rushes or reeds in or close to water, year-round. Often 'group breeds' with two or more females and at least two males.

Takahe *Porphyrio hochstetteri* E E
NOTORNIS, MOHO, SOUTH ISLAND TAKAHE
Previously widespread on both islands, now restricted to sanctuaries; Tiritiri Matangi, Kapiti, Mana and Maud islands.
Very large flightless bird (BL 63 cm) with massive red beak and frontal shield. Head neck and breast dark purplish-blue with brighter blue sheen. Back and tail darker with greenish sheen. Undertail white. Legs very stout and red. Builds a 'bower' in tussock grass. Very few, if any, left in wild, captive bred on rodent-free islands. SI: still found in small numbers in Murchison Mountains.

Australian Coot *Fulica atra* LC N
EURASIAN COOT, COMMON COOT
Widespread in NI and SI on lakes and lagoons.
All black bird (BL 35–39 cm) with white bill and frontal shield, legs dark. Scientific name translates as sooty black bird. Builds a floating nest in rushes or similar in lakes and ponds.

Banded Rail

Weka

Spotless Crake

Marsh Crake

Pukeko

Australian Coot

Takahe

WADERS

NATIVE BREEDING

Pied Oystercatcher *Haematopus finschi* **LC E**
TOREA, SOUTH ISLAND PIED OYSTERCATCHER
Widespread in inland SI in breeding season; spreads to NI coasts in winter.
Large, black-and-white wader (BL 46 cm), dark pink legs, long (80–90 mm) heavy red bill. Black above, white below, red eyering. White rump and white wingbar in flight. Sharp divide between black and white on breast, white extends onto shoulder. Nests in shallow scrape in riverbeds, lake shores and farmland to 1800 m.

Variable Oystercatcher *Haematopus unicolor* **LC E**
TOREA, TOREAPANGO, BLACK OYSTERCATCHER
Coastal beaches, especially northeast of NI and north of SI.
Large wader (BL 47–49 cm) with long (70–110 mm) heavy, red bill and pink legs. Black phase almost entirely black. Pied phase similar to *H. finschi*, but larger and less white on shoulder, wingbar and rump, and black–white divide on breast less defined. Intermediate phases are common. Nest is a shallow scrape on coastal beaches.

Spur-winged Plover *Vanellus miles* **LC N**
MASKED LAPWING, MASKED PLOVER
Widespread on all islands, especially wetlands and in open farming country.
Distinctive and noisy plover (BL 30–37 cm), upperparts mid-brown, black cap and shoulder band, otherwise white. In flight, wings broad and rounded, with broad black wing tips and trailing edges, spurs on elbows visible, tail black, rump white. Bill yellow with yellow facial 'mask', legs long, reddish. Flight irregular and undulating. Call, loud and varied including rattle and piping. Ground-nester in open country, near water. First recorded 1886.

Pied Stilt *Himantopus leucocephalus* **LC N**
POAKA, AUSTRALASIAN PIED STILT, BLACK-WINGED/WHITE-HEADED STILT, DOGBIRD
Widespread in wetlands, including farmland, on NI and SI.
Elegant, slim, black and white bird (BL 33–37 cm) with extremely long red legs (55–65 mm) and long, black, slender bill. Wings and back of neck black, otherwise white. Breeds on ground, loosely colonial, near water. Known to hybridise with the endemic Black Stilt.

Black Stilt *Himantopus novaezelandiae* **CE E**
KAKI
Central SI especially Twizel, in breeding season, scattered wetlands and riverbeds, NI and SI in winter.
Similar to but slightly larger (BL 40 cm) than the Pied Stilt, with which it sometimes hybridises, but all black, slight green sheen on back. Legs long, red. Bill long, fine, black. Juvenile has white head, neck and breast. Breeds in braided rivers in central Canterbury, migrates to coastal areas in winter. World's rarest wader with around 100 individuals, subject to an intensive recovery programme.

Wrybill *Anarhynchus frontalis* **V EG**
NGUTUPARORE, WRYBILLED PLOVER, CROCKBILL
Braided rivers in Canterbury and Otago in SI. Migrates to northern estuaries and harbours in winter.
Small dumpy plover (BL 20–21 cm). Head, nape and back uniform grey, undersides, throat and forehead white. Black throat band in breeding plumage. Bill long (26–30 mm), black, thin, and laterally curved to right. The only species in world with this feature. Breeds on ground in shingle in braided rivers.

Variable Oystercatcher Pied Oystercatcher

Pied Stilt Black Stilt

Spur-winged Plover

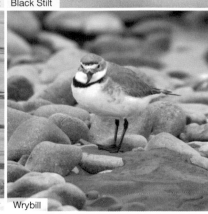
Wrybill

Banded Dotterel   *Charadrius bicinctus*   **LC**   **BE**
TUTURIWHATU, DOUBLE-BANDED PLOVER
Widespread on coastal beaches and inland braided rivers especially Canterbury. Some migration to coastal areas in the winter.
Small wader (BL 18–21 cm). Male bird brown above, white underparts. White forehead, chin, upper breast and body. Black band across throat, broad chestnut band on breast, which is lost out of breeding season. Female brown and white indistinct bands. Breeds on coastal beaches and in braided rivers.

New Zealand Dotterel   *Charadrius obscurus*   **E**   **E**
TUTURIWHATU, RED-BREASTED DOTTEREL
North of NI, mainly coastal, also Stewart Island mainly above tree line.
Two sub-species: Northern *C.o. aquilonius* and Southern *C.o. obscurus*. Small, rather bulky plover (BL 26–28 cm). Northern race brown back and wings, breast reddish buff, head brown with white eyebrow, forehead and chin, underwing greyish, undertail white. Loses buff breast in winter. Southern race larger with darker plumage, breast chestnut, similar in winter. Breeds on coastal beaches and dunes in NI; in cushion plants and rocks in subalpine areas in Stewart Is.

Black-fronted Dotterel   *Elseyornis melanops*   **LC**   **N**
BLACK-FRONTED PLOVER
Coastal rivers, NI and SI, especially Hawke's Bay and southern NI.
Similar to, but slightly smaller (BL 17 cm) than Banded Dotterel, with slightly longer legs, an orange red-bill and distinctive black forehead, eyestripe and chest band. Breeds riverbeds, gravel pits and open ground. Recent Australian arrival, first recorded breeding 1950s.

MIGRATORY
Plumage note: All of the waders that overwinter in New Zealand are in winter (non-breeding) plumage which makes identification difficult if not impossible. Their breeding plumage is generally very much more distinctive. The descriptions refer to winter or non-breeding plumage, except where indicated.

Pacific Golden Plover   *Pluvialis fulva*   **LC**   **NBM**
EASTERN, ASIATIC, LEAST OR LESSER GOLDEN PLOVER
Widespread Sept–Mar on or near coast, lagoons, harbours, estuaries; occasionally inland near pastures and wetlands on NI, SI and Chatham Is.
Medium-sized plover (BL 23–26 cm) with mottled brown and buff plumage, much darker on back, pale near-white throat, thighs and undertail. Bill short, dark, legs dark grey, light eyebrow. Breeding bird has dramatic black face, breast and undersides. Breeds in Siberia and Western Alaska.

Similar species: **Grey Plover** *P. squatarola* is larger, heavier and less common and lacks the buff colouration. In flight it has a black mark on underwing close to its body.

Bar-tailed Godwit   *Limosa lapponica*   **LC**   **NBM**
KUAKA, BAR-RUMPED, PACIFIC, OR SMALL GODWIT
Widespread, often in large flocks, coastal mudflats and lagoons; especially Kaipara and Manukau harbours, Firth of Thames, Bay of Plenty and Farewell Spit, Oct–April. Significant numbers remain in NZ over the winter.
Medium-sized (BL 37–39 cm), very long-billed (75–105 mm) wader. Head, neck and back mid grey-brown, throat and breast buff. Underparts off-white, some barring on flanks. Dark eyestripe, pale eyebrow. Bill dark tip, pinkish at base, slightly upturned; female noticeably larger. In flight rump and tail barred. Breeding plumage: head, breast and underparts rich chestnut, back also redder. Breeds in Siberia and Alaska. Individuals may complete southern migration in a single stage taking 8–10 days, the longest known non-stop flight by any bird.

Two other migratory godwits are also seen occasionally, the **Hudsonian Godwit** *Limosa haemastica* is similar to the Bar-tailed but has rather smoother grey-brown plumage, a more obvious eye-stripe and a white wingbar in flight; while the **Black-tailed Godwit** *L. limosa* is rather larger and has a longer, straight bill. Both have black tails visible in flight.

New Zealand Dotterel   Black-fronted Dotterel

Banded Dotterel   Pacific Golden Plover

Bar-tailed Godwit [Bar-tailed Godwit]

Two large uncommon migrants are the **Whimbrel** *Numenius phaeopus* and the **Eastern Curlew** *N. madagascariensis*. Both are large, long-legged birds, with mid-brown plumage and very long downcurved bills. *N. madagascarienis* is the larger (BL 63 cm) of the two and both are seen regularly in NI harbours, the Firth of Thames and on Farewell Spit.

Marsh Sandpiper *Tringa stagnatilis* LC NBM
Widespread but infrequent summer visitor, estuaries and lake margins.
Small (BL 22 cm), elegant, long-legged, pale grey wader. Upperparts pale brownish-grey, underparts white. In flight, white rump extends well up back, and barred tail is visible. Head pale grey-brown with white around bill and fait strip over eye. Bill long and finely pointed. Legs very long and dull or pale yellow. Breeding plumage: head, neck and upper breast flecked dark grey-brown, upperparts much darker. Significantly smaller than juvenile Pied Stilt, with which it is often seen. Breeds in Central Asia.

Wandering Tattler *Heteroscelus incana* LC NBM
AMERICAN, ASHEN OR TRINGINE SANDPIPER, *TRINGA INCANA*
Widespread but infrequent summer visitor, on rocky shores, generally solitary.
Medium-sized wader (BL 27 cm). Upperparts and breast mid-grey, underparts white. Bill long (35–40 mm), strong and black. Cap dark with white eyebrow. Legs strong, longish deep yellow. Breeding plumage, breast and flanks heavily barred. Clear whistling 6–10-note call. A very 'busy' bird. Breeds in Siberia, Alaska and Northern Canada.

Similar species: **Siberian Tattler** *H. brevipes* is very similar but is rather paler, especially the throat, and has a distinct white eyestripe. Its call is a high-pitched 'twheet twheet'.

Red Knot *Calidris canutus* LC NBM
HUAHOU, KNOT, LESSER, EASTERN OR COMMON KNOT
Sept–April, coastal wetlands throughout NZ, especially Kaipara and Manukau harbours, Firth of Thames and Farewell Spit. Some birds remain over summer.
A dumpy nondescript small wader (BL 23–25 cm) with dark, business-like bill (30–35 mm). Grey-flecked head and back, pale undersides, faint pale eyebrow. In flight, rump barred and whitish wingbar on edge of coverts. Legs greenish-grey. Breeding plumage: breast rich cinnamon, back flecked with chestnut. Breeds in Siberia.

Similar species: **Great Knot** *C. tenuirostris*, larger with longer bill, white rump visible in flight.

Curlew Sandpiper *Calidris ferruginea* LC NBM
PYGMY CURLEW, CURLEW STINT, REDCROP
Widespread summer visitor, small numbers. Estuaries and coastal wetlands in NI and SI.
Small, nondescript wader (BL 18–23 cm) with long (32–43 mm), thin, black, downcurved bill. Pale brownish-grey above, undersides white, faint white eyebrow, legs dark grey. Breeding plumage: rich chestnut colouration with horizontal barring on breast and abdomen. Breeds in Siberia.

Similar species: **Terek Sandpiper** *Xenus cinereus* is slightly browner on the back and has a long, thin, slightly upturned bill, orange-yellow at base, darker at tip and a white eyebrow. It has dull yellow-orange legs.

Red-necked Stint *Calidris ruficollis* LC NBM
Common but localized migrant, harbours and estuaries.
Very small (BL 15 cm) wader with short dark legs and short, dark, pointed bill. Back and wings pale grey-brown with dark primaries and secondaries with white wingbar visible in flight. Underparts and throat white. Faint pale eyebrow. Breeding plumage: rich chestnut face and neck, browner on back and wings. Breeds in Siberia and Alaska.

Ruddy Turnstone *Arenaria interpres* LC NBM
TURNSTONE, BEACHBIRD, CALICO-BIRD, SEA DOTTEREL
Widespread, coastal wetlands, Sept–Apr.
Stocky quite short-legged wader (BL 22–24 cm). Mid grey-brown flecked back and head. Throat and upper breast black-brown. Pale eyebrow. Bill short and black. Legs orange. In flight distinctive dark/white wing and tail pattern. Breeding plumage: dramatic chestnut and black. Breeds in Siberia, Alaska and Arctic Canada.

Eastern Curlew

Marsh Sandpiper

Wandering Tattler

Red Knot Curlew Sandpiper

Red-necked Stint Ruddy Turnstone

LAND BIRDS
KIWIS

There are five species of kiwi, three 'brown ones' and two 'spotted ones'. All are very similar physiologically and differ mainly in size and colouration.

The kiwi is a large, flightless, nocturnal bird with a very long downcurved bill with nostrils close to the tip. Brown Kiwis are just that, though greyer around head and neck. Spotted Kiwis are mainly grey with horizontal bands of brown, grey and off-white feathers, giving a spotted appearance. Legs are short and muscular with large, heavily toed feet. Feathers are more like hair than normal bird feathers.

Long sensitive rictal hairs at the base of the bill probably act as sensors for feeding at night. The bill may take up to five years to be fully grown.

Kiwis are found in areas of native bush from sea level to 1000 m. All kiwis are ground-nesting, in burrows, hollow logs or in dense vegetation. Eggs are extremely large at 440–530 g, which is 18–25% of the female's body weight, and are laid up to four weeks apart. Brown Kiwi lay June–December, Spotted Kiwi July–December (Great) or September–January (Little). Incubation is mainly carried out by males and takes two to three months. Young leave the nest after a few days, foraging independently. They are fully grown after 18 months.

South Island Brown Kiwi *Apteryx australis* **V EF**
TOKOEKA
South and west of SI, and especially Stewart Is.
The most widespread and numerous of the kiwis, there are considered to be four races or genetically distinct populations of the South Island Brown Kiwi: Haast, North and South Fiordland and Stewart Island. Stewart Island birds are generally larger than the mainland ones. They are 40–50 cm long, and weigh 2.2 kg (M) and 2.8 kg (F), with bills 8.1–12 cm long in males and 11.1–15.7 cm long in females.

North Island Brown Kiwi *Apteryx mantelli* **E EF**
ROWI
Northland, Coromandel and Bay of Plenty, also Te Urewera, Whanganui, Tongariro and Taranaki National Parks.
Virtually identical to the South Island Brown Kiwi, the North Island Brown Kiwi are distinguished only by location. Populations are recovering where there is active pest control.

Okarito Brown Kiwi *Apteryx rowi* **CE EF**
ROWI
Okarito, Westland.
Most closely related to North Island Brown Kiwi (similar size), it has slightly greyer plumage than other brown kiwis and has a very small population of less than 300. It is found only in the vicinity of Okarito in Westland.

Great Spotted Kiwi *Apteryx haastii* **V EF**
ROROA, ROA, GREAT GREY KIWI
Northwest of SI, Nelson, Buller and Huruni rivers to Arthur's Pass.
The largest kiwi (BL 45–55 cm; weight male 2.4 kg, female 3.3 kg), found mainly in higher altitude forest and tussock grasslands from Kahurangi National Park to Arthur's Pass.

Little Spotted Kiwi *Apteryx owenii* **V EF**
KIWI-PUKUPUKU
Several sanctuaries, including Kapiti, Tiritiri Matangi and Long islands; Karori Sanctuary, Wellington.
Smallest of the kiwis (BL 35–45 cm; weight M: 1150 g, F: 1325 g; bill M: 6.2–7.3 cm, F: 7.5–9.4 cm) and found in the wild only on offshore islands, otherwise in mainland sanctuaries.

South Island Brown Kiwi

North Island Brown Kiwi

Okarito Brown Kiwi

Great Spotted Kiwi

Little Spotted Kiwi

HAWKS AND OWLS

Australasian Harrier *Circus approximans* **LC** **N**
KAHU, HARRIER HAWK, SWAMP HARRIER (AUSTRALIA)

Widespread throughout NZ, especially open areas, wetlands, scrublands and farmland.
Large hawk (BL 50–60 cm; WS 1.2–1.45 m) wings long with broad-fingered tips, held in shallow 'V', long, banded, slightly rounded tail. Overall brown, darker above. Legs and eye yellow. Immature darker with dark eye. Female significantly larger than male. Often feeds on road carrion. Breeds on ground, in swamp, raupo or scrub, preceded by aerial display. One of the few native species to benefit from the destruction of the native bush and development of agriculture.

New Zealand Falcon *Falco novaeseelandiae* **NT** **E**
KAREAREA, BUSH HAWK, QUAIL HAWK

Largely forest dwellers. Widespread from Waikato southwards, apart from Stewart Is.
Small, fast falcon (BL 40–50 cm; WS 60–80 cm). Largely brown, dark flecked on wings and back. Throat and breast pale with dark flecking, thighs and undertail chestnut. Beak short, sharply hooked, yellow facial skin at base. Pale eyebrow, dark drooping 'whisker' below eye. Underwing barred. Legs yellow. Breeds on ledges of cliffs, under logs or amongst epiphytes.

Nankeen Kestrel *Falco cenchroides* **LC** **OV (Australia)**
AUSTRALIAN KESTREL (AUSTRALIA), HOVERER, MOSQUITO HAWK, WINDHOVER

Widespread but irregular visitor, open country.
Small raptor (BL 30–35 cm; WS 60–80 cm), often seen hovering over pasture or roadside verge. Male upperwing chestnut with large black tips, underwing largely white. Tail long, grey with black and white tip. Head grey, face white with dark 'tear' mark. Female largely reddish-brown above, buff breast and white undertail. Face buff with dark 'tear' mark. Legs yellow. Breeds in Australia, not recorded breeding in NZ.

Little Owl *Athene noctua* **LC** **I (Europe)**
GERMAN OWL

Farmland, SI especially in north and east.
Very small mottled owl (BL 21–23 cm). Plumage greyish-brown with white flecking and mottling, eyes large and yellow, tail short, heavy legs. Nocturnal but known to hunt in daytime. Breeds in holes or burrows in banks, buildings or trees.

Morepork *Ninox novaeseelandiae* **LC** **N**
RURU, BOOBOOK, SOUTHERN BOOBOOK (AUSTRALIA)

Throughout NZ especially in or near bush, often in urban areas.
Small, nocturnal, dark brown owl (BL 30–35 cm) with distinctive 'more pork' call. Dark brown on back with some light flecking, undersides lighter and barred. Eyes yellow, tail long and barred. Often observed in daytime, motionless on a dead tree or treefern. Nests in epiphytes or in a hole or fork of a tree.

Barn Owl *Tyto alba* **LC** **OV**

Breeds Australia, occasional visitor to NZ, most likely seen in open country and forest margins in NI and SI.
Medium-sized (BL 34 cm; WS 90–100 cm), very pale, largely nocturnal owl. Wings, back, tail and head yellowish-brown, body and facial disc white. A bird of open country and forest margins, appears almost all white when seen in car headlights.

Australasian Harrier

Australasian Harrier

New Zealand Falcon

Little Owl

Nankeen Kestrel

Barn Owl

Morepork

PARROTS

Kakapo *Strigops habroptilus* **CE EG**
GROUND, NIGHT OR OWL PARROT

Previously widespread, now confined to offshore islands, Codfish, Chalky, Anchor islands. About 120 birds survive.

Remarkable bulky, flightless, nocturnal parrot (BL 58–64 cm). Wings and back moss green, undersides yellow-green, tail bronze-green. Head, cap green, face has owl-like disk of pale hair-like feathers. Overall streaked and mottled with black and yellow. Large, powerful hooked beak. Distinctive booming call. Very complex lek mating system: males set up display area using loud booming call to attract females. Breeds opportunistically every 3–5 years dependant upon main food, the fruit of the Rimu tree. The breeding cycle lasts 4–5 months, starting late January.

Kea *Nestor notabilis* **V EG**
MOUNTAIN PARROT

SI only, high-altitude forest and alpine areas to 2100 m.

Large, broad-winged parrot (BL 45–50 cm). Head and body olive green, upperwing mid- green scapulars, primaries dark with blue or blue-green tinge. Underwing coverts bright orange-red. Bill has very long downcurved upper mandible. Noisy and acrobatic in flight. Very inquisitive, known to damage vehicles, camping and tramping gear. Polygamous, nests in holes or cavities in mountainous regions; males feed females during incubation and after hatching.

Kaka *Nestor meridionalis* **E EG**
BROWN OR BUSH PARROT

Two subspecies: NI *septentronalis* in central forests and Coromandel, also offshore islands. In SI *meridionalis*, in forested mountainous areas, also on Stewart Island and offshore islands.

Large, stocky, broad-winged forest parrot (BL 38–44 cm). Dark olive brown with red on flanks, rump, neck and underwing. Cap grey, yellowish area on face, long downcurved bill. Breeds in boles and holes of mature trees, male feeds female during incubation.

Eastern Rosella *Platycercus eximius* **LC I (Australia)**
COMMON, RED, OR RED-HEADED ROSELLA

NI only, Auckland area northwards and in Wairarapa and Hutt Valley.

Small, very brightly coloured, long-tailed parrot (BL 28–33 cm). Head neck and breast bright red, chin white, abdomen yellow to yellow-green, undertail red. Back and wings yellow scalloped with black, upperwing pale to dark blue. Tail green with pale blue outer feathers. Breeds in holes in trees, male feeds female during incubation.

Yellow-crowned Parakeet *Cyanoramphus auriceps* **LC E**
KAKARIKI

Widespread but scarce in podocarp and beech forests, also on offshore islands. Subspecies *forbesi* on Chatham Is.

Small, long-tailed forest parrot (BL 23–27 cm). Overall yellow-green, darker on back. Forecrown rich yellow, bright red frontal band. Small red patch on either side of abdomen at base of tail. In flight, outer primaries and coverts bright blue. Breeds in holes in trees, opportunistic, male feeds female during incubation.

Orange-fronted Parakeet *C. malherbi* (**CE E**). Very similar but with pale yellow forecrown and orange frontal band, found only in South Branch Hurunui and Hawdon rivers in North Canterbury and Maud Island.

Red-crowned Parakeet *Cyanoramphus novaezelandiae* **LC E (subspecies)**
KAKARIKI, PORETE, RED-FRONTED PARAKEET

Scarce in forested areas NI and SI, more visible on offshore islands.

Small yellowish-green forest parrot (BL 21–27 cm), almost identical to Yellow-crowned but forecrown red not yellow, red side patches more obvious, larger and heavier. Breeds in holes in trees, also in burrows in banks or dense vegetation, male feeds female during incubation.

Two other introduced parrots may be seen, often near urban areas. These are the all-white **Sulphur-crested Cockatoo** *Cacatua galerita* and the rather smaller pink and grey **Galah** *Cacatua roseicapilla*.

Kakapo Kea

Eastern Rosella Kaka

Yellow-crowned Parakeet Red-crowned Parakeet

PIGEONS, CUCKOOS, KINGFISHER AND SWALLOW

Kereru *Hemiphaga novaeseelandiae* NT E
KUKUPA, PAREA, NEW ZEALAND OR WOOD PIGEON
Widespread on all islands, in or near bush. Subspecies, *chathamensis,* or Parea, found in Chatham Is.
Only native pigeon. A large (BL 50 cm), handsome, heavy-bodied, broad-winged forest bird. Head, neck and upper breast, metallic green with bronze sheen. Back and wings purplish; wingtips dark with green sheen. Tail long with square black end. Lower breast, abdomen and undertail white. Bill, eye ring and feet red. Largest flighted bird in forest. Nests in trees, builds flimsy platform up to 15 m from ground.

Three introduced pigeons are found in New Zealand. The largest and most widespread is the **Rock** or **Feral Pigeon** *Columba livia* which has very variable plumage from light to dark grey with greens and purple on neck. It is much smaller than the Kereru and is not often found in forest habitats. Two other species are both smaller and paler/browner than the Rock Pigeon. The **Spotted Dove** *Stigmatopelia chinensis* is largely pinkish-brown with cinnamon breast, grey head, mottled chestnut back and a distinctive, broad, white-spotted, incomplete, black collar. While the **Barbary Dove** *Streptopelia roseogrisea* is almost uniform pinkish-beige with a much smaller partial black collar.

Shining Cuckoo *Chrysococcyx lucidus* LC BM
PIPIWHARAROA, SHINING BRONZE-CUCKOO (AUSTRALIA)
Widespread throughout NZ, shrublands, bush and suburbia. Overwinters in Bismarck Archipelago and Solomon Islands.
Very small, short-tailed cuckoo (BL 13–18 cm). Back, nape and crown bright metallic green, face, throat and undersides white with black barring. Sexes similar. Parasitic breeder, usually lays eggs in nest of Grey Warbler.

Long-tailed Cuckoo *Eudynamys taitensis* LC BM
KOEKOEA, KAWEKAWEA, LONG-TAILED KOEL (AUSTRALIA)
Mainly in bush from Hamilton south including Stewart Is., especially central NI and west of SI.
Large, long-tailed cuckoo (BL 40–42 cm), upperparts dark brown, barred red-brown. Face and breast pale but heavily streaked with brown. Bill large, light brown. Call a loud harsh 'schweesht', more often heard than seen. Parasitic breeder on Whitehead, Brown Creeper and Yellowhead.

At least four other cuckoos are occasional or regular visitors from Australia. Largest is the huge **Channel-billed Cuckoo** *Scythrops novaehollandiae* which is generally grey in colour with a grey head, red eye patch and massive peak bill. Much smaller is the **Pallid Cuckoo** *Cuculus pallidus*, which is all grey with black barring on the undertail, and the **Fan-tailed Cuckoo** *Cacomantis flabelliformis*, which is smaller still with grey back, cinnamon breast and strikingly barred black and white tail. Lastly, the **Oriental Cuckoo** *Cuculus saturatus* is all grey with black and white barring on the breast and undertail, it also has a brown phase.

New Zealand Kingfisher *Todiramphus sanctus* LC N
KOTARE, KINGFISHER, SACRED KINGFISHER *HALCYON SANCTA*
Widespread, though fewer in far south, especially in farmland, estuaries, mangroves.
Small, bright, large-billed bird (BL 21–24 cm). Often seen on exposed branches or wires. Crown, mantle and wing coverts green, rump, tail and flight feathers bright blue, neck and collar white, broad dark eyestripe. Bill very large, dark and heavy. NZ residents are subspecies *vagans*. Nests in holes in trees or banks, beside lakes and rivers, also road cuttings and coastal cliffs.

Welcome Swallow *Hirundo neoxena* LC N (partial migrants)
HOUSE OR PACIFIC SWALLOW
Widespread throughout NZ, apart from Fiordland and the alpine regions. Recent arrival, first breeding record 1958.
Small, streamlined bird (BL 15 cm) with deeply forked tail. Head, back, rump, tail and upperwing glossy blue-black. Underparts off-white, forehead, face and throat reddish-cinnamon. Mud nest attached to vertical surface or on beam in barn or similar. Up to three broods per season. Partial migrants with some birds moving north during winter.

Kereru

Rock Pigeon

Shining Cuckoo

Long-tailed Cuckoo

New Zealand Kingfisher

Welcome Swallow

NATIVE PASSERINES

Passerines is a term used to decribe 'perching' birds. In this section the native passerines are presented in approximate size order, beginning with the largest. The introduced passerines are described, along with game birds (which are all introduced birds), at the end of the Birds chapter.

Kokako *Callaeas cinerea* **E** **EG**
WATTLEBIRD, BLUE-WATTLED CROW
NI, mainly northern Urewera and Bay of Plenty, also on Little Barrier, Tiritiri Matangi and Kapiti islands. SI, probably extinct, possibly a few on Stewart Is.
Large, long-legged, long-tailed rather furtive forest bird (BL 38 cm). Plumage bluish-grey apart from black facial mask around eye and distinctive blue (NI) or orange (SI) wattles on either side of bill which is short and heavy. Song a haunting flute-like 'ko..ka..ko', mainly around dawn. Not easy to observe as they spend most of their time hopping through the forest canopy. Two subspecies: NI *wilsoni*; SI *cinerea*; latter possibly extinct, but a recently accepted 2007 sighting suggests that it may still survive in very small numbers. Nests in dense vegetation or in canopy, female builds nest, and incubates. Territorial.

Tui *Prosthemadera novaeseelandiae* **LC** **EG**
PARSON BIRD
Widespread in lowland forests and in suburban areas on all islands.
Large thrush-like bird (BL 27–32 cm) with broad, rounded wings. Plumage dark but with range of blue and green-bronze iridescent feathers and plumes, especially in breeding season. Sexes similar. Distinctive white 'bow tie' on throat, white wing patches and long black downcurved bill. Voice is a remarkable collection of melodious and unbird-like noises, whistles, cackles and gurgles. May make distinctive wing noise in flight. Immatures much less colourful and lack white 'bow-tie'. Subspecies *chathamensis* on Chathams is significantly larger. Nests in fork in canopy or sub-canopy tree, female builds and incubates. Very territorial in breeding season.

Saddleback *Philesturnus carunculatus* **NT** **EG**
TIEKE, JACKBIRD
Found only on protected, rodent-free islands off both NI and SI: Kapiti, Tiritiri Matangi, Hen and Little Barrier islands.
Medium-sized thrush-like forest bird (BL 25 cm). Sexes similar. Largely glossy black with rust brown 'saddle', rump and tail coverts. Bill long, very pointed and slightly downcurved. Conspicuous pendulous red wattles from base of gape. Immature much browner, lacks wattles and has whitish stripe at base of bill. Very active ground feeder so susceptible to alien mammalian pest species. Two subspecies: NI *rufusater*; SI and Stewart Is. *carunculatus*. Nests in holes in trees, in tree-fern crowns and dense epiphytic vegetation; female incubates. Territorial.

Stitchbird *Notiomystis cincta* **V** **EG**
HIHI
Mainly offshore islands: Kapiti, Little Barrier, Tiritiri Matangi, also Karori Sanctuary, Wellington, Maungatautari and the Waitakere Ranges near Auckland.
Small dark forest bird (BL 18–20 cm). Male black head, neck, breast and upper back, white blaze behind eye and rufous collar between black and generally mid-brown body, wings and tail. Female all mid-brown. In flight white wingbar in both sexes. Can be very noisy in groups. Very susceptible to alien mammal predators. Nests in hole high up in a tree. Several broods in a season. On Kapiti Island, Hihi are polyandrous, but not elsewhere.

Bellbird *Anthornis melanura* **LC** **EG**
KORIMAKO, MAKOMAKO, MOCKER
Widespread in native forest up to 1200 m all three main islands.
Small forest bird (BL 17–20 cm) with melodious voice. Largely olive green with near black flight feathers and tail which is clearly notched. Head, neck and upper breast dark olive with purple sheen, bill, long, thin, slightly downcurved, red eyering. Undertail light olive-buff. Female lighter with white moustache under eye. Four subspecies: NI, SI and Stewart and Auckland islands *melanura*; Three Kings *obscura*; Poor Knights *oneho*; Chatham *melanocephala* extinct. Nests in tree foliage, and occasionally in rock crevices. Female builds nest and incubates. Two clutches are common.

Tui Kokako

Saddleback Stitchbird (male)

Bellbird (male) Stitchbird (female)

New Zealand Pipit *Anthus novaeseelandiae* LC E (subspecies)
PIHOIHOI, AUSTRALASIAN PIPIT, RICHARD'S PIPIT
Widespread in open country on all three main islands from coast to alpine zone.
Small ground-feeding bird (BL 19 cm) with pale brown, flecked with darker brown plumage, white belly and undertail and outer tail feathers. White eyestripe and pale chin. Will often run and then fly a short distance. Four subspecies: NI, SI and Stewart Is. *novaeseelandiae*; Auckland and Campbell islands *aucklandicus*; Chatham Is. *chathamensis*, Antipodes Is. *steindachneri*. Nests on ground in grass tussock or similar; female builds and incubates. Strongly territorial.

Fernbird *Bowdleria punctata* LC EG
MATATA
Widespread but patchy distribution in low scrubby and swampy habitats, not forest.
Small warbler (BL 18 cm) with extremely long, rather shaggy tail. Overall reddish-brown, darker and heavily streaked above, much paler and less streaked below. Head, chestnut with white eyebrow, beak short and sharp. Skulking, rarely flies any distance. Not easily observed so the call, a double metallic 'shoo-click', can be diagnostic. Five subspecies: NI *vealeae*; SI *punctata*; Stewart Is. *stewartiana*; Codfish Is. *wilsoni*; Snares Is. *caudata*. Nests on ground in thick tussocky grass or rushes, two or three broods in a season on main islands.

New Zealand Robin *Petroica australis* LC E
TOUTOUWAI, BUSH ROBIN
Widespread but patchy, native bush, mainly central NI and northwest and southwest SI.
Small, perky, 'upright' forest bird (BL 18 cm). Three subspecies: NI *longipes*; SI *australis*; Stewart Island *rakiura*. NI: male almost all near-black, white spot on forehead and pale grey abdomen. Legs long and black, beak short and pointed. Female and juvenile rather greyer with light patches on breast. SI: male similar but abdomen and breast pale yellow with clear dividing line on breast and female all dark grey with yellowish belly. Nests in trees, often using old nests of other species. Female builds nest; female, fed by male, very territorial. Several broods in a season.

Similar species: **Black Robin** *Petroica traversi* is smaller, all black and found only in Chatham Islands. Rescued from extinction in 1981 when population reduced to five individuals and one breeding female 'Old Blue'. Population now around 250.

Whitehead *Mohoua albicilla* LC EG
POPOKATEA, BUSH CANARY
NI only, widespread in native bush south of Hamilton. Also on Little Barrier, Tiritiri Matangi, Kapiti and Mokoia islands.
Small, pale, long-tailed passerine (BL 15 cm). Male distinctive white head, neck and breast, back and tail mid-brown, undersides light brown, darker on flanks. Bill and eye black. Female similar but crown and nape pale brown. Nests in forest canopy or shrubs.

Yellowhead *Mohoua ochrocephala* E EG
MOHUA, BUSH CANARY, YELLOW-HEADED FLYCATCHER
SI only, beech forests in Fiordland, Caitlins and Arthur's Pass.
Small, colourful, noisy, gregarious, long-tailed passerine (BL 15 cm). Very similar to Whitehead, but bright yellow with brown back and tail. Much heavier. Scarce due to introduced predators. Nests in holes in beech trees; female builds nest and incubates.

Brown Creeper *Mohoua novaeseelandiae* LC EG
PIPIPI, NEW ZEALAND TITMOUSE
Widespread, forest and scrublands, coast to treeline, scarce in east, SI and Stewart Is. only.
Small, long-tailed, noisy, gregarious warbler (BL 13 cm). Upperparts reddish-brown, nape and face grey with fine white eyestripe behind eye. Chin and throat white, undersides buff. Nests in canopy, sometimes lower.

Fernbird

New Zealand Pipit

New Zealand Robin, South Island female

Whitehead (male)

New Zealand Robin, Stewart Island male

Yellowhead

Brown Creeper

LAND BIRDS – NATIVE PASSERINES

Tomtit *Petroica macrocephala* **LC E**
MIROMIRO, PIED TIT, NGIRU-NGIRU, YELLOW-BREASTED TIT
Widespread in forest and second growth manuka/kanuka scrub throughout country.
Small, dumpy forest bird (BL 13 cm). NI male upperparts, head, neck and upper breast black; body and undertail white, white wingbar and outer tail feathers. Bill very short and pointed, white spot on forehead, legs yellow. NI and SI female upperparts all mid-brown, white spot on forehead, throat grey, breast buff. SI subspecies similar but male has lemon-yellow breast becoming near orange at upper margin. Five subspecies: NI *toitoi*; SI *macrocephala*; Chatham Is. *chathamensis*; Snares Is. *dannefaerdi*; Auckland Is. *marrineri*. Snares Is. subspecies all black. Nests in hole in tree or thick vegetation. Female incubates, up to three broods in a season.

Silvereye *Zosterops lateralis* **LC N**
TAUHOU, WAXEYE, WHITE-EYE
Widespread throughout country, lowlands to treeline.
Very small olive-green bird (BL 12 cm) with conspicuous white eye ring. Back grey, undersides off-white, flanks cinnamon. Chin yellowish, undertail white. Often travels in small parties. Immatures lack eye ring. First recorded in NZ in 1832, breeding probably 1856. Suspended nest of grasses and root fibres.

Grey Warbler *Gerygone igata* **LC E**
RIRORIRO, GREY GERYGONE
Widespread throughout country, coast to subalpine zone; well adapted to suburban living.
In Chathams replaced by *G. albofrontata* Chatham Island Warbler, very similar but larger.
Small inconspicuous bird (BL 10 cm) of scrub forest. Upperparts uniform brownish-grey, tail darker with distinctive white tip. Underside almost white but with pale grey breast and throat. Pale eyebrow, short fine dark beak. Often hovers while feeding. Globular nest sometimes hanging, built by female. Normally two broods. Second clutch often host to Shining Cuckoo.

New Zealand Fantail *Rhipidura fuliginosa* **LC N**
PIWAKAWAKA, GREY FANTAIL
Widespread on all islands especially forest edges and scrubland.
Very small with extremely long tail (BL 16 cm, including 8 cm tail). Two colour morphs: pied commoner, back brown, head grey with white eyebrow, white and back bands on throat, breast and underside light cinnamon. Tail, two central feathers black, five outer feathers white. Black morph largely restricted to South, all black with white spot behind eye. Irregular bobbing flight using fanned-out tail while catching insects, frequently follows people, possibly as they may disturb insects. Three subspecies: NI *placabilis*; SI *fuliginosa*; Chatham Is. *penita*. Nests in thick scrub or tree fern. May start breeding as young as two months.

Rock Wren *Xenicus gilviventris* **V EG**
FIORDLAND ROCK WREN
SI only. Alpine and subalpine areas from northwest Nelson to Fiordland, Cobb Valley, Mount Arthur, Arthur's Pass and Homer Tunnel, Murchison Mountains. Irregular distribution, especially near screes and rock falls.
Very small, short-tailed, alpine bird (BL 9.5 cm), larger and with longer legs and shorter bill than Rifleman. Male olive green above, pale buff below, cream eyebrow, black patch on bend of wing, pale cinnamon patch on flanks. Female duller and browner than male. Nests in hollows, crevices or rock ledges; two clutches only if first fails.

Rifleman *Acanthisitta chloris* **LC EG**
TITIPOUNAMU, RIFLEMAN WREN
Widespread, locally common in beech and podocarps forest, NI and SI. Often in small parties out of breeding season.
Minute, short-tailed forest bird (BL 7–9 cm). Male head, neck and back and rump, bright olive green; underparts off-white. Upperwing, scapulars as back, but flight feathers dark. Distinct white eyestripe, bill long dark and slightly upturned. Female browner and with streaked plumage. Call very high-pitched 'zeep'. Birds in NI are subspecies *granti*, in SI subspecies *chloris*. Nests in holes in trees in lowland forest; often two clutches in a season.

Tomtit South Island subspecies (male)

Silvereye

Grey Warbler

New Zealand Fantail (with black morph, inset)

Rock Wren (male)

Rifleman (female)

Rifleman (male)

GAME BIRDS

Common Pheasant *Phasianus colchicus* **LC** **I**
RING-NECKED PHEASANT
Widespread in NI and Nelson, Canterbury and Otago in SI.
Large, very long-tailed bird (BL 55–100 cm, including tail 25–60 cm), male largely red-brown, head and neck dark glossy with bright red wattles. White neck ring not always present. Female pale to mid-brown, mottled, tail shorter. Breeds in ground in thick vegetation.

Indian Peafowl *Pavo cristatus* **LC** **I (Asia)**
COMMON PEAFOWL, PEACOCK (MALE)
Warmer and drier areas of NI, and northern SI.
Very large (BL 2.0–2.5 m including tail 0.8–1.6 m) distinctive bird. The male or Peacock has a ridiculously large, long, colourful tail. The female or Peahen is smaller (90 cm long) and largely brown and white. Not easy to mistake for anything else! A ground-nester, but no information on breeding in NZ.

Wild Turkey *Meleagris gallopavo* **LC** **I (North America)**
GOBBLER, FERAL TURKEY
Widespread in rough farm country, especially in NI.
Very large, long-legged, long-tailed, upright bird (BL 0.9–1.25 m). Largely black, male head and neck bare grey skin with distinctive large, upright red wattle on neck. Female grey skin on head only, no wattle. A ground-nester, but no information on breeding in NZ.

Helmeted Guineafowl *Numida meleagris* **LC** **I (Africa)**
TUFTED GUINEAFOWL
Mainly NI, agricultural land in Wanganui, Rotorua, Waikato and Northland.
Large, grey, rather dumpy feral game bird (BL 60 cm). Plumage dark grey all over with white flecking. White, bare skin, cheek patch, red wattles and distinctive bony crest on head. A ground-nester but no information on NZ habits.

Chukar *Alectoris chukar* **LC** **I (Asia)**
CHUKOR
Marlborough to Canterbury, mainly in the high country.
Medium-sized chunky partridge (BL 32–34 cm). Back wings, head, neck and rump grey-brown, throat cream with distinctive black border running from forehead and through eye. Flanks barred white, chestnut, black. Legs strong and red, bill red. Breeds on ground in rough grass or tussock. Nests Sept–Mar, young leave nest soon after hatching.

Brown Quail *Coturnix ypsilophora* **LC** **I (Australia)**
RAT QUAIL, SOMBRE, SORDID, SWAMP OR TASMANIAN QUAIL
NI north of Taupo, but mainly Northland.
Very small all-brown quail (BL 17–22 cm) with dark chestnut and black mottling on back, barred on undersides. Breeds on ground in rough grass or bracken, nests Sept–Jan, young leave nest soon after hatching.

California Quail *Lophortyx californicus* **LC** **I (North America)**
PLUMED QUAIL
Widespread apart from far south and southwest of SI.
Small quail (BL 25 cm) with distinctive forward-curving crest. Male brown with black face and throat, white eyebrow, brown cap, grey breast, speckled abdomen and barred underparts. Female largely mid-brown with speckled abdomen, some streaking on neck. Breeds on ground in rough grass near cover. Nests Sept–Feb, young leave nest soon after hatching.

Common Pheasant   Wild Turkey

Indian Peafowl (male)   Helmeted Guineafowl

Chukar   Brown Quail

California Quail (male)   California Quail (female)

INTRODUCED PASSERINES

Rook *Corvus frugilegus* **LC** **I (Europe)**
Mainly NI, Hawke's Bay and south to Wellington, also a few in Canterbury.
Large, all-black bird (BL 45 cm) with large pale grey bill, darker at tip, face bare and whitish. Nests in small colonies, top of tall trees; female incubates.

Australian Magpie *Gymnorhina tibicen* **LC** **I (Australia)**
WHITE-BACKED MAGPIE, BLACK-BACKED MAGPIE
Widespread in NI, and east of SI, especially farmland.
Large, unmistakeable, black and white crow (BL 45 cm). Two subspecies: White-backed *G.t. hypoleuca* predominates but Black-backed in Hawke's Bay and Northern Otago. Former has white nape extending onto back, while latter has white nape only. Bill pale grey with dark tip. Legs long and grey. Call is an uncharacteristic flute-like whistle. Nests in small colonies, high up in trees; female incubates.

Blackbird *Turdus merula* **LC** **I (Europe)**
Widespread in all habitats except thick bush, but especially parks and gardens, to 1400 m.
Medium-sized (BL 25 cm) bird. Male all-black with bright orange bill and eyering, legs, dull orange. Female all mid-brown, bill dull orange with darker tip. Nests in thick shrubs or hedges, sometimes in old sheds; female builds and incubates. May have two broods.

Common Myna *Acridotheres tristis* **LC** **I (India)**
NI only, open country and urban areas mainly north of Volcanic Plateau.
Stocky brown bird (BL 24 cm) with near-black head, orange-yellow bill, eye golden yellow, yellow ear patch. White wing patches and undertail. Tail black, rounded with white tip. Nests in holes in tree, cliff or building, incubation mainly by female.

Song Thrush *Turdus philomelos* **LC** **I (Europe)**
THRUSH
Widespread especially farmland, parks and gardens.
Medium-sized bird (BL 23 cm), all mid-brown on back, breast buff with distinctive brown flecking, undertail near-white. Bill dark brown with distinctive orange-yellow gape. Female builds mud-lined nest in a small tree or dense shrub, and incubates. May have two broods.

Starling *Sturnus vulgaris* **LC** **I (Europe)**
Widespread in open country to 1200 m, especially farmland and urban areas.
Medium-sized, short-tailed, short-winged bird (BL 21 cm). Overall dark with white spotting and green sheen on wing. Bill long and sharp. Breeding plumage dark purple sheen on head and breast, bill yellow. Nests in holes in tree, cliff or building. Two broods common.

Skylark *Alauda arvensis* **LC** **I (Europe)**
Widespread but mainly in open grasslands, agricultural areas, especially drier areas to the east.
Small light-brown bird (BL 19 cm), heavily streaked and striated, very pale undersides, slightly darker breast. Pale eyebrow and distinct crest that is raised at will. Male hovers high up, often out of sight of naked eye, singing continuously. Nests on ground in open grassland, built by female who also incubates.

Yellowhammer *Emberiza citrinella* **LC** **I (Europe)**
Widespread in open country to 1600 m, and in parks and gardens throughout NZ. Tends to form flocks in autumn and winter.
Distinctive long-tailed yellow bunting (BL 16 cm). Male upperparts streaked yellow-brown, rump rich chestnut, tail dark, white outer feathers, slightly forked. Head bright yellow with longitudinal brown stripes on crown. Undersides yellow with indistinct cinnamon breast band. Female similar but yellow less bright and breast band dull grey-brown. Song 'little bit of bread and no… cheese'. Nests in scrub, shrubs, hedges or long grass; two broods common.

Australian Magpie   Rook

Common Myna

Blackbird

Starling

Yellowhammer

Song Thrush   Skylark

Cirl Bunting *Emberiza cirlus* **LC** **I (Europe)**
Widespread mainly Nelson to south Otago, open country, coastal grasslands. Forms flocks in autumn and winter, especially in coastal grasslands.
Similar to Yellowhammer (BL 16 cm) but less colourful, browner and streakier. Breeding male has yellow face with black eyestripe and black 'beard' that joins with eyestripe, also rufous on flanks and wings. Nests in thick low vegetation, gorse or similar, female incubates. Two broods common.

Chaffinch *Fringilla coelebs* **LC** **I (Europe)**
Widespread throughout NZ, coast to subalpine, especially farmland and suburbia. Often flocks in autumn and winter.
Small colourful finch (BL 15 cm). Male undersides soft cinnamon-peach becoming richer on face, head and nape blue-grey with black forehead; mantle chestnut, rump grey-green, tail black with white outer feathers. Wings dark with distinctive white double wingbar. Female largely brown, pale eyebrow, back darker and double white wingbar. Bill short and conical, male heavier than female. Nests in trees or dense scrub, female incubates.

Greenfinch *Carduelis chloris* **LC** **I (Europe)**
GREEN LINNET
Widespread especially in farmland, forest edges and suburbs. Often forms flocks in autumn and winter.
Medium-sized (BL 15 cm), stocky finch. Overall greenish-yellow rather browner on back and with bright yellow-green on forepart of wings and sides of base of tail. Tail black and noticeably forked. Male generally brighter than female. Bill pale, short and pointed. Nests in trees or larger shrubs, female incubates. Normally two broods in season.

Dunnock *Prunella modularis* **LC** **I (Europe)**
HEDGE SPARROW
Widespread but not conspicuous, scrubland, orchards, parks and gardens.
Small (BL 14 cm), rather quiet, almost furtive brown bird. Head and back dark brown, grey collar and breast. Undersides grey with brown flanks. Tail brown. Pale eyebrow. Nest built by female in thick undergrowth and incubated by her. Normally two or three broods in a season.

House Sparrow *Passer domesticus* **LC** **I (Europe)**
Widespread throughout NZ, especially farmland and urban areas.
Small brown gregarious finch, (BL 14 cm). Male back dark red-brown streaked with black. Crown grey, nape chestnut, cheeks and underside light grey-brown, black bib extending to breast in breeding season. Bill short and conical. Female all brown, pale below, darker with streaking above, brown cap, pale eyestripe. In flight both show short white wingbar. Domed nest in hole in tree, cliff, building or similar.

Goldfinch *Carduelis carduelis* **LC** **I (Europe)**
Widespread throughout NZ, especially farmland, forest borders and suburbs. Often flocks in winter.
Small brightly coloured finch (BL 13 cm). Back light brown, rump white, tail black with white spots near end. Abdomen, light brown and white. Head black and white with distinctive red face, bill short and pointed, wings black and golden yellow. In flight yellow wingbar obvious. Nests in small trees or shrubs, females incubate. Male feeds female and chicks during brood period.

Lesser Redpoll *Carduelis cabaret* **LC** **I (Europe)**
Widespread throughout NZ, especially south of Hamilton. Often flocks in autumn and winter.
Small, (BL 12 cm) brown, heavily streaked finch; darker on back, paler below, undertail near-white. Red on forecrown and black chin, pale eyebrow and faint wingbar. Male develops crimson breast in breeding plumage. Very short pointed bill. Nests in low shrub, female builds nest and incubates.

Cirl Bunting Chaffinch

Dunnock

Greenfinch

House Sparrow

Lesser Redpoll Goldfinch

MAMMALS

The conservation and species status abbreviations for marine and land mammals are as for birds, see page 29.

MARINE MAMMALS

In contrast to the paucity of native land mammals, and in line with its extensive coastline, marine mammals are well represented in New Zealand. They fall into three groups; whales, dolphins and pinnipeds or seals.

WHALES

While many species occur from time to time in New Zealand waters, three species are most commonly seen. The **Humpback Whale** *Megaptera novaeangliae* is up to 16 m long and has a very knobbly head and extremely long flippers; it is well known for making spectacular leaps from the water, their tail flukes are individually identifiable and they have a minimal dorsal fin. They migrate past New Zealand, heading north up the east coast in spring and south down the west coast in autumn.

The **Southern Right Whale** *Eubalaena australis*, up to 18 m long, is slightly larger than the Humpback, and often has clearly visible clumps of barnacles attached to its head. Named because they were the 'right' whale to catch, they breed in sheltered coastal waters on Campbell and Auckland islands and are sometimes seen around the coast of the main islands.

The most easily viewed whale is the **Sperm Whale** *Physeter macrocephalus*, up to 18 m in length, and identifiable by its square snout and forward pointing blow, they, too, have a minimal dorsal fin and can be found year-round off Kaikoura on the east coast of South Island, where as deep divers, to 400 m, they are attracted by the submarine canyons, which are rich in marine life.

DOLPHINS

While many species of dolphin have been observed in New Zealand waters, six species are resident, ranging in size from the Orca, which can be up to 10 m in length, to the diminutive Hector's Dolphin at 1.5 m.

Orca *Orcinus orca* LC N
KILLER WHALE

Locally common in small family groups, often with distinct territories, around coast. Occasional beachings.

Large, to 10 m, black with distinctive white undersides, flank and neck markings. Large dorsal fin is also diagnostic; the fin of a large bull being nearly a metre tall. Orca live in family groups or pods and have often clearly defined ranges, frequently found close inshore where they may feed on flatfish and rays, also on squid, dolphins, seals and fur seals depending on location.

Long-finned Pilot Whale *Globicephala melas* LC N
BLACKFISH

Not often seen inshore, except when beached.

Large, up to 6 m, and all black with a distinctive hooked dorsal fin. Found in pods of up 120 animals, sadly quite frequently found beached in large numbers.

Common Bottlenose Dolphin *Tursiops truncatus* LC N
Northeast of the NI, the north of SI, with a small population in Fiordland.

Possibly the best known of the dolphins, a large (length to 4 m) steely-grey dolphin with distinctive long snout. Frequently plays in bow wave of boats. Found worldwide.

Short-beaked Common Dolphin *Delphinus delphis* LC N
Found mainly in Hauraki Gulf and east coast of NI, especially Bay of Plenty.

Significantly smaller (BL 2.1 m–2.6 m) than the Bottlenose Dolphin but with a noticeable beak and distinctive colouration, dark above but pale grey flanks and lighter belly. Can form schools of up to several thousand animals, will also play around moving boats, found in Pacific and Atlantic oceans and in the Caribbean and Mediterranean seas.

Southern Right Whale

Sperm Whale

Orca

Long-finned Pilot Whale

Humpback Whale

Common Bottlenose Dolphin

Short-beaked Common Dolphin

MARINE MAMMALS   **85**

Dusky Dolphin *Lagenorhynchus obscurus* **LC N**

A coastal species found on east coast, from East Cape southwards, major populations in Marlborough Sounds and off Kaikoura, where you can swim with them.

Medium-sized (BL to 2 m) dolphin with no beak but distinctive coloring; near black back, near white belly with white running up neck and becoming an eyestripe. Very acrobatic species.

Hector's Dolphin *Cephalorhynchus hectori* **V E**

Inshore waters SI, especially Banks Peninsula and west coast Karamea to Hokitika. Maui: NI west coast Dargaville to Taranaki.

Smallest of all dolphins (BL to 1.5 m). Mid grey above, pale grey belly with distinctive black facial markings and diagnostic large, round, black dorsal fin. Two subspecies: *C.h. hectori*, which is found around SI, and may itself be divided into two or three races, and the Maui Dolphin *C.h. maui*, which is critically endangered with a population of around 50 adult individuals. There is doubt that regulations brought in to protect it from set net fishing will be enough to halt its slide into extinction.

PINNIPEDS

Eight species of seal are recorded in New Zealand waters: five Phocids or true seals – **Leopard Seal** *Hydrurga leptonyx*, **Weddell Seal** *Leptonychotes weddellii*, **Crabeater Seal** *Lobodon carcinophagus*, **Ross Seal** *Ommatophoca rossi* and **Southern Elephant Seal**, all but the last being occasional visitors from the Antarctic; and three Otariids or 'eared' seals – the **New Zealand or Hooker's Sea Lion** and the **New Zealand** and **Sub-Antarctic Fur Seals**, the last, only an occasional visitor, is distinguished by its pale face mask and mohican crest on the males.

Southern Elephant Seal *Mirounga leonina* **LC N**

Rocky beaches south of Dunedin to Bluff.

Largest of the seals, identified by size; adult males are huge (M: BL 5 m/WT 3700 kg; F: BL 2.5 m/WT 800 kg) and have a very distinctive inflatable nose. Breeds on Antipodes and Campbell Islands and appears on the main islands occasionally.

New Zealand Fur Seal *Arctocephalus forsteri* **LC N**

KEKENO, AUSTRALASIAN OR SOUTHERN FUR SEAL

Widespread. Breeding on rocky beaches, mainly SI, Banks Peninsula, Fiordland, Kaikoura, Stewart Is, Farewell Spit. More widespread at other times, Cook Strait, Bay of Plenty.

Dark brown, thick-furred seal with up-pointed nose (M: BL to 2.5 m/WT to 160 kg; F: BL to 1.5 m/WT to 70 kg). Distinguished from Sub-Antarctic Fur Seal *A. tropicalis* by all-brown fur. Breeds November to January with mating place taking place immediately after birth of pup. Population increasing from very low level 100 years ago. Found also in Australia.

New Zealand Sea Lion *Phocarctos hookeri* **E N**

HOOKER'S SEA LION

Sandy beaches of the lower SI, from Dunedin to Bluff, also Stewart Is.

Significantly larger than the Fur Seal (M: BL to 3.5 m/WT to 450 kg; F: BL to 2 m/WT to 160 kg), and more heavily built with a blunt snout. Male dark brown with thick furry ruff over head and neck which is much lighter when dry. Female much smaller and paler, especially when dried out. Breeds in Auckland and Campbell Islands, with small groups on the sandy beaches of lower South Island and Stewart Island. Bulls can be very aggressive in the mating season, December to January. This population is threatened by commercial fisheries.

Dusky Dolphin   Hector's Dolphin

Bull Southern Elephant Seal

New Zealand Fur Seal

New Zealand Fur Seal

Bull New Zealand Sea Lion   New Zealand Sea Lion, young bull with females

NATIVE LAND MAMMALS

New Zealand currently has only two species of native land mammal, both bats; a third bat species became extinct in 1967. This is largely due to the New Zealand continent breaking away from Gondwana 70–80 mya, before the evolution and spread of many mammalian orders. The recent discovery of the fossil of a small mouse-like mammal in Otago has both indicated that non-flying land mammals did make it to New Zealand and perhaps also suggested that the timing of the separation of New Zealand from Gondwana was before the development of more 'sophisticated' modern mammals.

Long-tailed Bat *Chalinolobus tuberculatus* **V E**
Widespread both NI and SI in mature forest.

Very small (BL 6–9 cm, of which the tail is 3.5 cm, and WT 8–11 g), dark red-brown to black forest bat, with small ears. The wings (WS 25–30 cm) extend to include the long, pointed tail. Feeds on flying insects. Roosts in large mature forest trees. Separate subspecies on NI and SI. Closely related to an Australian family of bats.

Lesser Short-tailed Bat *Mystacina tuberculata* **E EF**
Widespread throughout the country but restricted to a small number of known locations.

A most unusual bat, the only survivor of Mystacinidae family. Larger than the long-tailed bat (BL 6–9 cm, of which tail is 1 cm, and WT 11–25 g), grey-brown in colour and with large prominent ears, a long tubular nose, very short tail, large and well-developed feet and wings, (WS 25–30 cm) that fold into a leathery pouch, which it uses to walk on the forest floor and to climb trees. Mixed diet of insects, berries and nectar. Specific pollinator of a root parasite, *Dactylanthus taylori,* which produces large amounts of nectar on which the bat feeds. A poor flyer, slow and rarely above 3 m. A lek breeder, roosts in mature forest trees, often in quite large numbers. Three subspecies: Northern *aupourica* found in Northland and on Little Barrier Island; central *rhyacobia* is the most numerous and found in central NI and on Kapiti Island; southern *tuberculata* is restricted to northwest Nelson, Fiordland and Codfish Island. Very vulnerable to introduced predators due to ground foraging habits.

INTRODUCED/ALIEN LAND MAMMALS

There are more than 30 introduced mammals in New Zealand, most of them deliberate introductions. With few native predators, they have thrived and caused widespread ecological and environmental damage.

MARSUPIALS

Between 1858 and 1870 some twelve species of marsupial were introduced from Australia of which the **Brush-tailed Possum** *Trichosurus vulpecula* and five species of wallaby have survived.

The Possum was introduced for its fur, but it soon escaped. They are largely nocturnal and are estimated to number 70–90 million, spread over the entire country, eating around 140,000 tonnes of vegetation *every night*. They cause significant damage to many areas of native bush having a predilection for mistletoe *Elytranthe* spp., rata *Metrosideros* spp. and fuchsia *Fuschia excoriticata*. They also predate young and eggs of small birds as well as native invertebrates, and they can carry bovine tuberculosis. You are most likely to see them as roadkill.

The other five species of wallaby, only two are found on the main islands. The most widespread is the **Red-necked** or **Bush Wallaby** *Macropus rufogriseaus* a large pale grey or fawn wallaby with distinctive rufous neck and shoulders, found in tussock grasslands in South Canterbury, especially Hunter Hills and Grampian Mountains. The **Dama** or **Tammar Wallaby** *Macropus eugenii,* is small with silver-grey fur and a short, pointed tail. It is found on Kawau Island in the Hauraki Gulf, where it was introduced around the 1870s, and around Rotorua, Bay of Plenty and Waikato in the North Island. It is considered a pest as it feeds on and damages native vegetation and young forest plantings.

The other three species, **Black-tailed** or **Swamp Wallaby** *Wallabia bicolor*, the **White-throated** or **Parma Wallaby** *Macropus parma* and the **Brush-tailed Rock Wallaby** *Petrogale penicillata* are found only on Kawau Island in the Hauraki Gulf. A number of the White-throated Wallaby have been captured and returned to Australia where it is scarce.

INSECTIVORES

The **European Hedgehog** *Erinaceus europaeus* was introduced from Britain to control garden pests around 1870. Largely nocturnal, it has become a widespread pest species due to their predation on skinks, the eggs of ground-nesting birds and native invertebrates including Weta.

LAGOMORPHS – RABBITS AND HARES

The **European Rabbit** *Oryctolagus cuniculus* is the most common, widespread and damaging of the two species of lagomorph. It was introduced by the early settlers for food and skins, and by 1869 it was reported as a pest in Southland. Rabbits damage herbaceous vegetation, their burrowing causes erosion and they are an economic pest; 15 rabbits can each as much vegetation as one sheep.

The **European** or **Brown Hare** *Lepus europaeus*, introduced around 1850 for food and sport, is significantly larger and darker brown than the rabbit, with long black-tipped ears. It is not a serious pest but does harm young trees.

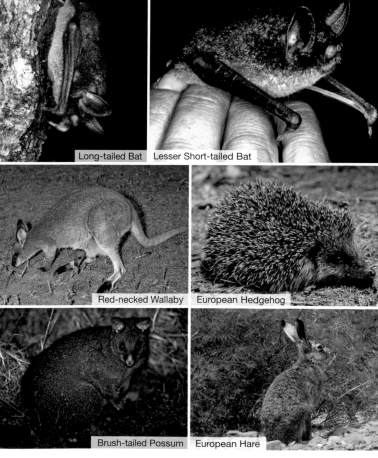

Long-tailed Bat   Lesser Short-tailed Bat

Red-necked Wallaby   European Hedgehog

Brush-tailed Possum   European Hare

CARNIVORES

The four introduced carnivores have a dramatic and damaging impact on native wildlife. The **Feral Cat** *Felis catus* is widespread, especially in remote areas, and while they do prey on rodents and mustelids, they are serious predators of birds and lizards. The other three carnivores are all mustelids, which were first introduced in 1867 to control rabbits. However, their principle food is invertebrates, mice and birds. The most widespread and numerous is the **Stoat** *Mustela erminea* (32–37 cm) a slim, long-tailed animal, dark brown with whitish underparts and a black tip to its tail. The **Weasel** *Mustela nivalis* (22–25 cm) is much smaller than the stoat with a shorter tail that lacks the black tip. The largest of the three is the **Ferret** *Mustela putorius furo* (47–60 cm) clearly distinguished by its light brown head with a dark brown band or mask across the face covering the eyes.

RODENTS

The **Kiore** or **Polynesian Rat** *Rattus exulans*, which arrived with early Polynesian settlers, was the first of four species of rodent to make it to New Zealand. Previously widespread, they are now restricted to Fiordland, Stewart Island and offshore islands, apparently driven out by the **House Mouse** *Mus musculus* and the **Brown** or **Norway Rat** *Rattus norvegicus* which is thought to have arrived with Captain Cook. The **Black** or **Ship Rat** *Rattus rattus* arrived later and was not widely distributed until the 1880s. The latter is darker and has a tail much longer than its body and large hairless ears. All four species are serious pests, creating havoc with the native fauna, particularly invertebrates, birds, frogs and reptiles. The most serious threat today appears to be from the Black Rat, which has brought local extinctions to nine species of native birds.

UNGULATES

Most of the eleven species of ungulate, or hooved mammals, were introduced for sport purposes with many becoming pests. Their overall impact on the ecosystem is extremely damaging, especially on native plant species. **Wild Pig** *Sus scrofa* and **Feral Goat** *Capra hircus* were first introduced by Captain Cook in 1773. They were also released on outlying islands as a source of food for shipwrecked sailors. Between them, pigs and goats can destroy an ecosystem, eating anything and everything.

Himalayan Tahr *Hemitragus jemlahicus*, also known as Himalayan Mountain Goat, and **Chamois** *Rupicapra rupicapra* were introduced in 1904 and 1907, respectively, as game animals. The latter were a gift from the Emperor Franz Josef. They are found at higher altitudes in the Southern Alps. Both are selective grazers, damaging native flora, especially vulnerable alpine and subalpine vegetation.

Red Deer *Cervus elaphus* introduced in 1851, is the commonest and most widespread of the seven species of deer introduced to New Zealand. With the rise in value of venison, wild deer were trapped and red deer farming is now widespread. The other deer species are **Sika** *Cervus nippon*, **White-tailed** or **Virginian** *Odocoileus virginianus*; **Fallow** *Dama dama*; **Wapiti** or **North American Elk** *Cervus canadensis*; **Sambar** *Cervus unicolour* and **Javan Rusa** *Cervus timorensis*. The damage done to the vegetation by deer is immense; they are selective grazers, preferring growing shoots and thus preventing seedlings from developing.

As a result of the problems posed by all these introduced and destructive pests, New Zealand has become the world leader in the elimination of invasive animal species, with experts proving advice and assistance to eradication programmes worldwide.

Ferret

Chamois

Himalayan Tahr

Feral Goat

Wild Pig

White-tailed Deer

Red Deer

AMPHIBIANS AND REPTILES

FROGS

The only native amphibians in New Zealand are four species of endemic frog. All very small, they lack an external eardrum, do not croak but emit very small squeaks, their ribs are not attached to their backbones. Three of them have no tadpole stage and lack webbed feet. They also lack the ability to catch prey by extending the tongue. All are largely nocturnal, very local and best viewed by torchlight on a really wet night in thick bush! Once more widely distributed, they have suffered from the introduction of rats and mustelids.

All except Hochstetter's Frog lay up to 20 small clusters of gelatinous eggs (up to 20 mm in diameter) in wet or muddy sites under rocks and logs, from which emerge fully formed froglets with short tails which are subsequently absorbed into the body.

Archey's Frog *Leiopelma archeyi* **CE EG**
Native bush and subalpine scrub to 1000 m. Restricted to Coromandel Peninsula and southwest Waikato.
A tiny brown frog (BL M to 31 mm; F to 37 mm) with a green sheen and black blotchy markings. Distinct ridge running back from eye. Feet unwebbed.

Hamilton's Frog *Leiopelma hamiltoni* **E EG**
Restricted to Stephens Island, a nature reserve in the Cook Strait. Some recently translocated to Chetwode Island in the Marlborough Sounds.
The largest of the native frogs (BL M to 43 mm; F to 49 mm), silvery brown, distinct eye ridge and with black blotchy markings. Feet unwebbed.

Maud Island Frog *Leiopelma pakeka* **V EG**
Moist coastal bush on Maud Island in Marlborough Sounds.
Slightly smaller than, but virtually indistinguishable from, Hamilton's, but found only on Maud Island (BL M to 43 mm; F to 47 mm).

Hochstetter's Frog *Leiopelma hochstetteri* **V EG**
Native bush close to streams up to 800 m from East Cape to Waikato and southern Northland.
Mid-brown with rather warty looking skin, small eye ridges and dark brown blotchy markings, feet webbed. The most aquatic and most likely native frog to be seen (BL M to 38 mm; F to 47 mm), and the only one to have a free-swimming tadpole stage.

INTRODUCED AUSTRALIAN FROGS

Southern Bell Frog *Litoria raniformis* **LC I (Australia)**
GOLDEN BELL FROG, GROWLING GRASS FROG
Widespread farmland, edge of bush, lakesides, swamps and streams NI and SI.
Back largely bright green with brownish markings, undersides light brown, inside of thighs blue (BL M to 65 mm; F to 92 mm). Lateral line running back from eye. Voice a series of sharp staccato croaks.

Green and Golden Bell Frog *Litoria aurea* **LC I (Australia)**
Habitat as for *L. litoria* but restricted to northern NI.
Similar to Southern Bell Frog but largely golden brown with bright green patches (BL to 85 mm). Toe pads noticeably wider than toes. Voice a long drawn out 'crooooaak'.

Whistling Frog *Litoria ewingii* **LC I (Australia)**
BROWN TREE FROG
Widespread especially in SI and Stewart Is., also west of NI, in moist habitats, even above tree line.
A small, all brown frog (BL M to 37 mm; F to 47 mm), darker on back than sides, with inside of thighs orange. Voice a cricket-like 'weeep weeeep weeeep'.

Archey's Frog Hamilton's Frog

Hochstetter's Frog

Maud Island Frog Green and Golden Bell Frog

Whistling Frog Southern Bell Frog

REPTILES

New Zealand has over 90 native species of reptile, divided into three groups, Geckos, Skinks and Tuatara. The abbreviation SVL is for Snout Vent Length, ie length excluding the tail.

Tuatara *Sphenodon punctatus* **E EO**

Previously widespread, now restricted to offshore island reserves, and mainland sanctuaries such as Karori and Mount Bruce.

Heavily-built reptile (Male SVL 310 mm, Tail to 300 mm, WT to 1.3 kg; Female much smaller to 500 g) with rough, mottled grey-green or grey skin and a distinctive soft-spined dorsal ridge.

Tuatara are truly amazing animals. They are the only surviving member of the order Sphenodontida (and have many features linking them to birds). All the other species of this order became extinct at least 60 million years ago, soon after the separation of New Zealand from Gondwana. Tuatara lack an external ear, have a third or parietal eye, and have three rows of teeth, the lower jaw teeth fit between the two sets on the upper jaw, and all are an integral part of the jawbone. Males have no external sexual organ.

Tuataras have a lifespan of at least 60 years in the wild, and more than 100 in captivity. They start breeding aged 10 or more, and females breed every 2 to 5 years. Eggs can take 16 months to hatch and the young can continue to grow for 35 years.

GECKOS

Geckos have rough, loose skin, often brightly coloured or patterned, with small dome-like scales that do not overlap. Their toes have soft pads with hairy ridges on the undersides, enabling them to climb very smooth surfaces. Their eyes have vertical pupils and no eyelids, but transparent scales covering the eyes that they clean with their tongue. When they shed their skin, they do it in one piece. They are viviparous.

There are seven genera—*Dactylocnemis, Hoplodactylus, Mokopirirakau, Naultinus, Toropuku, Tukutuku* and *Woodworthia*—all are viviparous and all vocalise. Currently 43 species are recognised.

Common Gecko *Woodworthia maculata* **LC EG**

Forest, scrub and grassland to 1700 m. Widespread on NI and north-west SI, especially coastal areas.

Medium-sized brown gecko (SVL to 8.2 cm) with varied dark and lighter brown patches, bands or stripes. Pale underneath. Inside of mouth pink.

Forest Gecko *Mokopirirakau granulatus* **LC EG**

Widespread in forest and scrub to 1400 m northern NI and north-west SI.

Large, well camouflaged, brown gecko (SVL to 8.9 mm) with varied patterns of dark and light browns, almost lichen or bark-like, may also be greenish. Inside mouth is bright orange.

Green Gecko *Naultinus elegans* **LC EG**

Widespread on NI excluding Northland.

Large, all green or all yellow gecko (SVL to 9.5 cm) with varied, irregular yellow markings, underside pale green, flanks on male may be light blue. Inside mouth blue, tail prehensile. Subspecies Wellington Green Gecko *N. e. punctatus* in south and east.

Jewelled Gecko *Naultinus gemmeus* **LC EG**

SI and Stewart Island only, widespread in forest, scrub and tussock to east of Southern Alps.

Distinctive smallish green gecko (SVL to 8 cm) with yellow stripes running full length of body. Canterbury males may be largely brown. Inside of mouth blue.

Tuatara

Common Gecko

Forest Gecko

Green Gecko

Jewelled Gecko

SKINKS

Skinks can best be distinguished from geckos by their sleek, shiny appearance, due to their smooth overlapping scales. They move quickly and have pointed toes which are smooth on the undersides. They have moveable lower eyelids which are transparent in some species, and round pupils. Skinks are less colourful than geckos, being mainly combinations of brown and black and they shed their skins bit by bit. All but one of the native skinks are viviparous, and most are diurnal. There are at least 48 species of skink in New Zealand, all but one endemic. The largest, *Oligosoma northlandi*, is now extinct thanks to alien mammals, while many other species are restricted to offshore islands or remote alpine areas.

Copper Skink *Oligosoma aeneum* **LC EG**

NI only, widespread in open areas with cover, common in gardens, and on coast down to high tide mark.
Smallest native skink (SVL to 6.2 cm, with tail 12 cm), coppery brown back, paler sides with dark irregular markings, belly pale.

Common Skink *Oligosoma polychroma* **LC EG**

This species complex is widespread in open areas with cover, river beds, coastal areas. Each of the taxa in the complex have restricted distributions.
Medium-sized, dark brown skink (SVL 7.7 cm) with prominent cream-coloured lateral stripes. Underside grey or pale yellow.

Shore Skink *Oligosoma smithii* **LC EG**

Northern NI and islands only.
A small (SVL 8 cm) skink found close to shoreline and in back dunes. Diurnal and often seen basking in the sun. Has a varied appearance heavily flecked with darker and lighter markings, and a wide variety of colour morphs from light brown to near black. The belly is pale cream to near orange with lighter flecking on the throat.

Ornate Skink *Oligosoma ornatum* **LC EG**

NI only; widespread bush and open areas with cover. Nocturnal and crepuscular, active at dawn and dusk.
Medium-sized (SVL to 8 cm), light brown back, cinnamon on flanks with black markings especially along colouration margin. Belly paler, tail may have pale blotches.

Scree Skink *Oligosoma waimatense* **E EG**

SI only; open rocky areas, riverbeds, scree and tussock to 1400 m, diurnal, sun bather.
Large (SVL to 10.7 cm), pale greyish brown on back and sides with variegated black patterns and longitudinal streaking on back.

Rainbow Skink *Lampropholis delicata* **LC I (Australia)**

DELICATE SKINK, GARDEN SKINK
Open areas from Palmerston North northwards, especially urban gardens and industrial sites.
Very small (SVL 45 mm, with tail 55 mm), short-legged skink with a long, thin tail. Overall a rich brown but with a narrow black stripe from snout, through eye and becoming a broad dark band along flank. Paler on belly. Oviparous, several females may use same nest site. An accidental introduction from Australia. Declared an unwanted organism in 2010.

Green Skink *Oligosoma chloronoton* **V EG**

Tussock grasslands to 1700 m in south Canterbury, Otago, Southland and Stewart Is.
A large (SVL to 125 mm) sometimes brightly coloured skink. Back varies from dark brown through pale brown, pale green, olive green to bright green, with pale, black-edged flecks. Back edges generally creamy brown, flanks quite varied, with dark and pale stripes. Often live in family groups. Flecking tends to be heavier the further south.

Other species: The critically endangered **Otago Skink**, or Mokomoko, *Oligosoma otagense* is a large (SVL 130 mm) strikingly patterned skink; body black or dark brown with heavy grey, green and yellow flecks. Restricted to Schist outcrops in Southern Otago. The **Grand Skink** *O. grande* is smaller (SVL 110 mm) and with similar but rather finer markings. It has a similar habitat and range.

Common Skink

Copper Skink Shore Skink

Scree Skink Ornate Skink

Rainbow Skink Green Skink

INVERTEBRATES

The conservation and species status abbreviations for invertebrates are as for birds, see page 29.

SLUGS, SNAILS AND VELVET WORMS

There are some 1400 species of native slug and snail in New Zealand, however unless you look carefully for them, you are most likely to see one of the 30 or so introduced species, notably the **Brown Garden Snail** *Cantareus aspersus* and the **Grey Field Slug** *Deroceras reticulatum*, both of which are well known to gardeners. The native species, with notable exceptions, are generally smaller and adapted to life on the forest floor, many of them are threatened due to loss of habitat and introduced mammalian predators.

Kauri Snail *Paryphanta* spp. **T EG**
PUPURANGI
Forest and native scrub, northern NI, and offshore pest-free islands.
Large, dark greenish-brown to black, flattened shell (60–80 mm across shell), carnivorous and cannibalistic. Nocturnal, lives in the leaf litter on the forest floor and feeds on earthworms and soft-bodied invertebrates. Young initially live in trees. Inactive during dry periods. Two species: *P. busbyi* and *P. watti*.

Giant Land Snail *Powelliphanta superba* **E EG**
PUPURANGI
Lowland and montane forest, central NI southwards. On SI, largely on western, moister side. Some species found in subalpine tussock.
Very large carnivorous snail (110 mm across shell, WT 90 g), inhabits the forest floor and feeds on earthworms and soft-bodied invertebrates. Shell flattened, mid to dark brown. Nocturnal. Young initially arboreal. Twenty-one species, some with patterned and reddish or yellow shells, all endemic, some very local.

Flax Snail *Placostylus* spp. **V EG**
PUPUHARAKEKE
Northland and offshore islands around northern NI.
Large mid to dark brown snail with conical shell (length of shell 85–115 mm), feeds on forest floor on leaves and detritus, does not feed on flax. Young initially live in trees. Three species.

Leaf-veined Slug *Pseudaneitea* spp. **V/T EG**
PUTOKO ROPIROPI
Damp places in broadleaved lowland bush on NI and SI.
There are some 30 species of native New Zealand slugs. All are well camouflaged with a leaf-vein pattern on their upper side and up to 40 mm in length. Thought to live mainly on algae and fungi on plant leaves and stems. They have only one pair of tentacles – introduced species have two. Native slugs do not damage garden plants.

Peripatus *Peripatoides novaezealandiae* **T/E EG**
NGAOKEOKE, VELVET WORM
Found in very damp habitats, especially rotting logs on NI, SI and Stewart Is.
Unusual invertebrate rather like a centipede but with 13–16 pairs of short, stumpy legs on a body up to 60 mm in length. Dark blue velvety appearance with gold spots, other species have different colouration. Carnivorous and cannibalistic, feeds on other invertebrates by ejecting a sticky glue which entraps them. A member of the family Onychophora which dates back more than 500 million years. Five named species in New Zealand, more recent discoveries are not yet described or named.

Kauri Snail   Flax Snail

Leaf-veined Slug   Giant Land Snail

Peripatus

SLUGS, SNAILS AND VELVET WORMS   **99**

SPIDERS

There are probably 3000 native species of spiders in New Zealand, of which only around 1100 have been fully described and identified. Many of these are endemic. There are also a number of introduced species. Spiders are widespread and are found in all habitats and ecosystems up to 3500 m. While most species are venomous to some degree, only a few are capable of inflicting a painful bite. The following are a small sample of arachnid types that you may encounter. LS = leg span BL = body length.

Nurseryweb Spider *Dolomedes minor* **LC N**
Widespread shrublands, also waste ground, gardens.
Large, (LS 60 mm BL 30 mm) fast-moving, pale-brown to grey spider with yellowish markings on cephalothorax. Carries egg sac until hatching then constructs a dense web at end of branch of tree or shrub, typically gorse or manuka. A similar species, the **Water Spider** *D. aquaticus* attaches its nursery web to stones in river or stream beds, this species hunts on surface of water.

Black Tunnelweb Spider *Porrhothele antipodiana* **LC N**
Widespread, especially lowlands, common in gardens.
Large, heavy-bodied spider (LS 40 mm BL 20–30 mm) with dark abdomen and legs and red-brown cephalothorax, spinerettes extend beyond tail. Builds silken tunnel and preys on beetles and other insects, snails and occasionally even mice. The inspiration for 'Shelob' in the film of 'Lord of the Rings'.

Garden Orbweb Spider *Eriophora pustulosa* **LC N**
Widespread, especially in gardens.
Small spider with large, rather spiky abdomen (LS to 25 mm BL to 12 mm), often quite colourful. Spins orb-shaped web to catch its prey.

Black-headed Jumping Spider *Trite planiceps* **LC N**
Widespread, especially lowlands, commonly found indoors.
Small, distinctively-shaped spider (LS 15 mm BL 40 mm), very long front legs and cephalothorax black, large pair of central eyes which it uses to stalk and leap onto prey. Abdomen pale brown, other legs reddish brown. One of 150 similar species found throughout the country.

Sheetweb Spider *Cambridgea* spp. **LC E**
Widespread NI, SI and Stewart Is., especially in native forest.
Very large, grey-brown, nocturnal bush spider (LS 200 mm). Spins sheet web up to 1 m across. Sometimes found in houses and gardens. 30 species, some much smaller.

Daddy Longlegs Spider *Pholcus phalangioides* **LC I**
Widespread and numerous, mainly indoors, not generally found south of Christchurch.
Very small spider with immensely long legs (LS to 200 mm BL 5–8mm), body almost translucent, often seen carrying eggs. Builds web to entrap flying insects.

Katipo *Latrodectus katipo* **LC E**
RED KATIPO
Coastal dunes, in grass, driftwood or flotsam and jetsam, from Northland as far south as Greymouth and Dunedin.
The only seriously venomous native species. A small (LS 32 mm BL 10 mm) black spider with a pea-size abdomen. Only the female can bite humans and it has a distinctive orange-red stripe with white surround on its back and a red hourglass mark underneath. Males and juveniles are smaller and black and white. Feeds on insects and other invertebrates.

Similar species: The **Redback** *Latrodectus hasselti*, from Australia, is very similar to the Katipo, and again only the female can inflict a bite to humans. Unlike the Katipo, however, it is found inland, around Wanaka on SI and New Plymouth on NI.

The **White Tailed Spiders** *Lampona cylindrata* and *L. murina* are also from Australia. Both are pirate spiders, feeding solely on other spiders. Dark grey with a distinctive white tip to the abdomen, they are widespread on NI and SI. Their bite can be painful and has been likened to a bee sting.

Black Tunnelweb Spider

Nurseryweb Spider

Garden Orbweb Spider

Black-headed Jumping Spider

Sheetweb Spider

Daddy Longlegs Spider

Katipo

BUTTERFLIES AND MOTHS – LEPIDOPTERA

Some 92% of native Lepidoptera are endemic.

IS IT A BUTTERFLY OR A MOTH?

	Butterfly	Moth
Antennae	Long and thin with clubbed tip	Often feathery, no club tip
Wings at rest	Normally folded upwards	Laid flat or along body at rest
Flight	Diurnal (during the day)	Largely nocturnal (at night)

BUTTERFLIES

New Zealand has relatively few species of native butterfly, though recent research has revealed that there are many more than originally thought, possibly as many as 70. There are also a small number of introduced species. This relative paucity is in part because butterflies generally inhabit open country while most of New Zealand was originally covered in thick bush.

Monarch Butterfly *Danaus plexippus* LC N
KAHUKU
Widespread in gardens in urban and suburban areas, NI and SI.
Largest butterfly in New Zealand (WS 85–90 mm), thought to have arrived in the late 19th century. Bright red-brown wings with black edges and veins, male has small dark scent pouches on vein next to abdomen. Back wings have large white markings on underwing edges. Dramatic black, white and yellow banded caterpillars feed mainly on Swan Plant *Gomphocarpus fructicosus*, a member of the milkweed family. A migratory species in its original range in the Americas, the New Zealand population is not known to migrate.

Red Admiral Butterfly *Vanessa gonerilla gonerilla* LC E
KAHUKURA, *BASSARIS GONERILLA GONERILLA*
Gardens and open country on NI and SI.
Large distinctive butterfly (WS 65 mm) with red bars on wings, forewing tips have white 'mirrors', rear have dark eyes with bluish centres. Rear wing bars have four white-centred dark 'eyes'. Black caterpillars feed on native Ongaonga *Urtica ferox* and the introduced Stinging Nettle *Urtica incisa*. Hibernates during winter.

Yellow Admiral Butterfly *Vanessa itea* LC N
KAHUKOWHAI, *BASSARIS ITEA*
Widespread in open country and gardens NI and SI.
Smaller than Red Admiral (WS 40–45 mm) and with yellow patches and white mirrors on forewings, but not on rear wings. Frequently seen on *Buddleia* bushes. Black caterpillars feed on Ongaonga *Urtica ferox* and Stinging Nettle *Urtica incisa*. Hibernates in winter.

Australian Painted Lady Butterfly *Vanessa kershawi* LC NBM
PEPE PARAHUA
Widespread in open lowland areas and gardens NI and SI. Mainly summer.
Mottled orange and brown (WS 45–50 mm) with dark forewing tips with white 'mirrors', rear wings have four eyes. Underwing mottled browns and greys. Often feeds on *Buddleia*, caterpillars feed on plants of the daisy family. Regular summer visitor from Australia, has bred here on occasion.

Monarch Butterfly, male

Yellow Admiral Butterfly

Monarch Butterfly, female

Red Admiral Butterfly

Australian Painted Lady Butterfly

Tussock Ringlet   *Agyrophenga janitae*   **LC E**
JANITA'S TUSSOCK
High tussock grasslands, SI only.
Medium-sized brown butterfly (WS 25–40 mm) with orange-brown patches, containing white-centred dark-brown spots in the middle of each wing. Green caterpillar feeds on Snow Tussock and other grasses.

Similar species: **Common Tussock** *A. antipodum*, which lacks the white spots, and **Harris' Tussock** *A. harrisi*, which is restricted to the Nelson area.

Black Mountain Ringlet Butterfly   *Percnodaimon merula*   **V EG**
PEPE POURI
SI tussock lands, altitude 1000–2000 m.
Dark brown butterfly (WS 40–45 mm) with four coalescing, near-black, white-centred eyes on forewing tips. Caterpillar feeds on Blue Tussock and other grasses.

Long-tailed Blue Butterfly   *Lampides boeticus*   **LC N**
Widespread on NI and northern tip of SI.
Small brownish butterfly (WS 30 mm) with mauve sheen, small tail on each rear wing. First recorded in New Zealand in 1965. Caterpillars feed on flowers of Gorse, Broom, and also Peas and other legumes.

Common Blue Butterfly   *Zizina labradus labradus*   **V N**
PEPE AOURI
Widespread in gardens and farmland on NI and northwest of SI below 1000 m.
Very small (WS 25 mm) mauve-blue butterfly with brown edges to wings. Small green caterpillars feed on clover, lucerne and similar legumes.

Southern Blue Butterfly   *Zizina labradus oxleyi*   **V N**
SI scrublands to east of the Southern Alps.
Similar but smaller (WS 20 mm) than Common Blue, with which it sometimes hybridises, and with dark zigzag band on underside of wings.

Common Copper Butterfly   *Lycaena salustius*   **LC N**
PEPE PARARIKI
Widespread from coast to tussock lands on NI and SI.
Distinctive, small, red-orange butterfly (WS 33–35 mm) with dark brown markings especially round edges of wings. SI variety much heavier markings than NI variety. Recent research shows considerable number of species. Green caterpillar feeds on Pohuehue leaves.

Boulder Copper Butterfly   *Boldenaria boldenarum*   **LC N**
Mountains and tussock lands, especially on SI.
Very small butterfly (WS M 20–22 mm F 23–25 mm). Male is iridescent blue and smaller than copper-coloured female. Recent research indicates there are several distinct species.

Cabbage White Butterfly   *Pieris rapae*   **IP**
PEPEMA, WHITE BUTTERFLY
Widespread, especially suburban areas and farmland.
Medium-sized white butterfly (WS 45 mm) with variable dark spots on wings and dark forewing tip. Caterpillar feeds on cabbage and other brassicas, also nasturtium. Introduced 1930.

Tussock Ringlet

Black Mountain Ringlet Butterfly

Long-tailed Blue Butterfly

Common Blue Butterfly

Common Copper Butterfly

Southern Blue Butterfly

Cabbage White Butterfly

Boulder Copper Butterfly (female)

MOTHS

New Zealand has a large and varied native moth population, with over 1700 species (94% endemic). Over half the species are micro-moths, with a wingspan of less than 20 mm.

Gum Emperor Moth *Opodiptera eucalypti* I (Australia)

Agricultural and urban areas, from Nelson northwards.

Very large mid to light brown moth (WS to 120 mm) with prominent 'eyes' on all four wings. Female has large body and small antennae, male has slim body and large feathery antennae. Short-lived, adults have no mouth parts and are unable to feed. Caterpillars green with light green lateral stripe and red tufts. Feeds on eucalyptus, silver birch, apricot and grape vine.

Puriri Moth *Aenetus virescens* V E

PEPETUNA

Native bush in NI, often attracted to lights.

New Zealand's largest moth (WS 130–150 mm). Forewings are green with brown hieroglyphics. Caterpillars develop under bark of mature trees and take up to seven years to mature, growing up to 120 mm in length.

Cabbage Tree Moth *Epiphryne verriculata* LC EG

PURERE TI

Widespread where there are cabbage trees.

Long-winged moth (WS 60–65 mm). Wings covered with fine lateral pale and darker brown stripes giving it camouflage when at rest on dead cabbage tree leaves. Small greenish caterpillars feed at night on cabbage tree leaves.

Silver Y Moth *Chrysodeixis eriosoma* LC N

Widespread, especially in gardens and cultivated areas.

Small nocturnal moth (WS 35–40 mm), mid-brown, folds wings back on body giving an inverted 'V' shape, distinctive silvery 'Y' markings on forewings. Green caterpillars eat Poroporo, Rengarenga and garden vegetable leaves.

Cinnabar Moth *Tyria jacobaeae* I (Europe)

Widespread in farmland, chiefly lower NI and upper SI.

Small, bright crimson, day-flying moth (WS 33–35 mm). Forewings darker than rear wings. Caterpillars dramatic yellow and black striped; feed on Ragwort *Senecio jacobaeae*. Introduced 1929 in an unsuccessful attempt to control this noxious pasture weed.

Magpie Moth *Nyctemera annulata* LC E

PURERE URI

Widespread at lower altitudes.

Distinctive moth (WS 40 mm), wings are dark brown with large white spots or blotches, body rather wasp-like with yellow and black bands. Day-flying. Caterpillar is woolly and dark with longitudinal yellow stripes, feeds on Ragwort and Cineraria, poisonous to most predators. Very similar to Australian species *N. amica*, with which it is thought to interbreed.

LOOPER MOTHS

There are nearly 300 species of looper moths in New Zealand, mostly brown with an amazing variety of patterns, resulting in many being known as carpet moths. Looper moths normally rest with wings outspread, though often with hind wings covered by forewings.

Lichen Moth *Declana atronivea* (NI) *D. egregia* (SI) LC E

PUREREHUA, ZEBRA OR ZEBRA LICHEN MOTH

Widespread in native and regenerating bush, NI and SI.

Forewings strongly patterned off-white and brown to resemble lichen (WS: NI 40–42 mm SI 50–55 mm). Caterpillars are twig-like and feed on Five Finger and lichens.

Kumara Moth *Agrius convolvuli* V N

HIHUE, CONVOLVULUS HAWK MOTH

Widespread near cultivated areas, especially early evening.

Large fast-flying moth (WS 100–110 mm) with grey forewings, brownish rear wings and banded body with red segment near head. Caterpillar strikingly marked brown with curved spine at rear end and eyes along sides.

Gum Emperor Moth

Cabbage Tree Moth

Silver Y Moth

Puriri Moth

Cinnabar Moth

Magpie Moth

Lichen Moth

Kumara Moth

BEETLES – COLEOPTERA

Beetles are the largest order of insects with 370,000 species worldwide, of which some 5500 are found in New Zealand; 90% of them endemic, some are flightless. Beetles have two pairs of wings but the forewings form a hard rigid protective cover over the hind wings and abdomen. Beetles are very varied in habitat being found from the coast to 3700 m, and even around hot springs. Most are found in native bush and alpine vegetation, they feed on a wide range of foods including fungi, plants and other animals.

LONGHORN BEETLES – CERAMBYCIDAE

There are some 180 species of Longhorn Beetles in New Zealand, all characterised by long, narrow bodies and long, at times very long, antenna. Adults are vegetarian, the grubs, which bore into dead wood and can spend many years before metamorphosing, are considered a delicacy in many parts of the world. Some species are flightless and many emit a noise on being touched.

The largest New Zealand species is the **Huhu** or **Tunga Rere** *Prionoplus reticularis*. Dark brown with paler longitudinal stripes, it is about 50 mm in length, does not eat and lives for only about two weeks as an adult. Commoner is the introduced **Burnt Pine Longhorn** *Arhopalus tristis* which is dark brown and has fairly short antenna. Similar but paler brown and with longer antennae is the native **Kanuka Longhorn** *Ochrocydus huttoni*.

SCARAB BEETLES – SCARABAEIDAE

A large group with some 140 species in New Zealand, most of which are endemics, including chafers and dung beetles. All have roundish, generally shiny head and wing covers, brown or black, though chafers are mostly green. Adults feed mainly on vegetable matter and dung. The grubs, which are widely consumed worldwide as food, are curled as in the letter 'C'.

The **Large Sand Scarab** or **Mumutawa** *Pericoptus truncatus* measuring 30–35 mm and the **Small Sand Scarab** *Pericoptus* spp. are coastal species found in sand dunes and under driftwood from Otago northwards. More widespread is the small **Black Scarab** *Heteronychus arator* and the bright green **Mumu Chafer** *Stethaspis longicornis*.

LADYBIRDS OR MUMUTAWA – COCCINELLIDAE

A family of very small, 2–5 mm long, often brightly coloured and patterned beetles. There are some 40 species in New Zealand, over 50% native and many endemic. Both grubs and adults feed on powdery mildews, aphids, mealybugs, scale insects and mites. The grubs do not look grub-like, and have legs as long as the adults. Ladybirds are often used as a biological control of plant pests.

Common in New Zealand are the introduced **Two-spotted Ladybird** *Adalia bipunctata* and **Eleven-spotted Ladybird** *Harmonia conformis*, both are red with black spots; also the **Steel-blue Ladybird** *Halmus chalybeus* which is just that. Native species include the **Orange-spotted Ladybird** *Coccinella leonina* which has orange spots on a black background, **Fungus-eating Ladybird** *Illeis galbula* which is black and yellow.

Huhu Beetle   Burnt Pine Longhorn   Kanuka Longhorn

Large Sand Scarab   Black Scarab

Mumu Chafer   Two-spotted Ladybird

Eleven-spotted Ladybird   Steel-blue Ladybird   Orange-spotted Ladybird

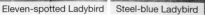

WEEVILS – CURCULIONIDAE

Members of the largest animal family in the world, with over 48,000 species, of which some 1540 are found in New Zealand. They are generally small, the largest being 25–30 mm long, with a long 'snout' and antenna which have a distinct bend or elbow in them. All are vegetarian, feeding on plants, often seen as agricultural pests, especially the introduced species. Many of the native flightless species are eaten by rodents and are threatened or endangered.

The endemic **Helm's Beech Weevil** *Anagotus helmsi* is found from Coromandel to Stewart Island, the grubs feed on all 5 species of native beech. The **Flax Weevil** *Anagotus fairburni* is a large, flightless brown weevil generally restricted to islands and some subalpine areas, where they are not predated on by rats. It is found only on its host plant flax *Phormium* spp. The endemic **Speargrass Weevil** *Lyperobius huttoni* is also flightless and is found in tussock grasslands in the South Island from 1200–1600 m. The male of the remarkable **Giraffe Weevil** or **Tuwhaipapa** *Lasiorhynchus barbicornis* can be up to 70 mm in length, half of which is an elongated snout with antennae at the tip. The female is much shorter with a snout shorter than the male (but still long). It is found in native bush from Canterbury northwards, the grubs take two years to mature and leave a square hole on departing.

Both grubs and adults of the **Gorse Seed Weevil** *Exapion ulicis* feed on gorse seeds. It was introduced to help control Gorse, which it has singularly failed to do as it has only one generation per year and the gorse flowers twice.

GROUND BEETLES – CARABIDAE

Another large family with 29,000 species worldwide and some 540 species in New Zealand, most of which are endemic and flightless. They are generally shiny black and can best be found under logs, stones and general detritus on the forest floor.

The **Metallic Ground Beetle** *Megadromus antarcticus* is 25–30 mm long, and is found largely in Canterbury in bush, farmland and gardens. The 30 mm-long **Stinking Ground Beetle** *Plocamostethus planiusculus* is one of many similar species and is found from Coromandel southwards, it feeds on insects, can bite and emits a foul smell if touched. Much smaller, at 15 mm, is the **Cosmopolitan Ground Beetle** *Laemostenus complanatus*, an introduced species which is found throughout the country under stones, woodpiles and the like in farmland and urban areas.

STAG BEETLES – LUCANIDAE

A smaller family of some 1300 species, 50 in New Zealand, mostly endemics and some flightless. Stag beetles are often distinguished by their very large jaws, especially in the males. Either black or brown, with elbowed antennae, the adults feed on fruit juices or sap. The grubs, which are 'C' shaped, spend up to 6 years in rotting wood.

The largest stag beetle in New Zealand, at 42 mm long and 20 mm wide, is the **New Zealand Giant** or **Helm's Stag Beetle** *Geodorcus helmsi*, found in lowland rainforest in south Westland and Fiordland. More widespread and smaller at 25 mm is **Earl's Stag Beetle** *Dendroblax earlii* which is often seen flying on spring or summer evenings. The grubs live in the soil and feed on plant roots.

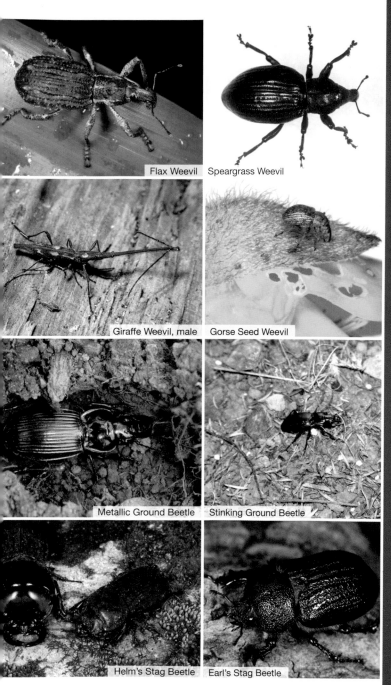

Flax Weevil   Speargrass Weevil

Giraffe Weevil, male   Gorse Seed Weevil

Metallic Ground Beetle   Stinking Ground Beetle

Helm's Stag Beetle   Earl's Stag Beetle

WETAS, GRASSHOPPERS AND CRICKETS – ORTHOPTERA

Orthoptera, which translates as 'straight winged', have rigid forewings and long hind legs enabling them to jump large distances. Males of many species 'sing' by rubbing their wings against each other or their legs. Most are flightless. Females have large sabre-shaped ovipositors. There are more than 130 species in New Zealand.

WETAS – ANOSTOSTOMATIDAE

Very widespread and numerous, at least 110 species are known, all endemic, all are nocturnal and flightless.

Giant Weta   *Deinacrida* spp.   **E EG**
WETAPUNGA
Formerly widespread now very restricted; Little Barrier, Mahurangi and other island sanctuaries.
Enormous dark brown insect (BL to 100 mm and weight to 70 g), with extremely long antenna at 160 mm. Back legs 90–100 mm long and heavily spined, shield over neck is wider than head. All nine known species are very susceptible to predation by introduced rodents.

Previously widespread, now confined to offshore islands apart from the **Scree Weta** *D. connectens,* which is widespread on the alpine scree slopes of SI.

Tree Weta   *Hemideina* spp.   **LC EG**
PUTANGATANGA
Widespread and common in bush, orchards and gardens.
Much slimmer than Giant Weta (BL 70 mm), banded dark and light brown abdomen, shorter legs and antenna. Back legs heavily spined and raised up in a defensive posture when alarmed. Omnivorous and with formidable jaws. Seven known species.

Ground Weta   *Hemiandrus* spp.   **T EG**
Widespread from coastal sand dunes to alpine areas.
Much smaller than the Tree Weta but with very long antenna (BL 25–30 mm, antenna 50–60 mm). It has similar dark and light brown banded colouration, omnivorous though lacking large jaws. Their back legs have no large spines. It lives in holes in the ground but forages above ground and in trees and shrubs. More than 30 species, only 7 of which have been described.

The **Mountain Stone Weta** *H. maori* is found is tussock grasslands on SI and can withstand being frozen solid.

Cave Weta   *Gymnoplectron* ssp., *Pharmacus* spp. & *Neonetus* ssp.   **LC EG**
Found in caves, disused building and under rocks and logs throughout the country from sea level to the alpine zone.
There are well over 50 species of Cave Weta, all endemic. They are dark or light brown, soft-bodied, and vary in size from 15–50 mm long. They have impressively long legs, antenna up to four times the body length and are wingless and deaf.

Giant Weta

Tree Weta   Ground Weta

Mountain Stone Weta   Cave Weta

GRASSHOPPERS

Two groups are found in New Zealand: Short-horned Grasshoppers Acrididae, most of which are flightless, abound in the alpine areas, especially in the South Island; and Long-horned Grasshoppers or Katydids, which are more common in the North Island. Each species has a distinct song. Females are significantly larger than males.

SHORT-HORNED GRASSHOPPERS – ACRIDIDAE
KOWHITIWHITI

There are some 15 known species and all but one are flightless and endemic. The exception, and the largest in New Zealand, is the **Migratory Locust** or **Kapakapa** *Locusta migratoria*, a worldwide species which is up to 55 mm long. It is found on both North and South islands, but does not form swarms as in hotter climates. Short-horned grasshoppers are widespread in the South Island, mainly at higher altitudes up to 2000 m but also in coastal tussock and grassland.

Two common upland species are the **Northern Tussock Grasshopper** *Paprides nitidus* and the **Southern Tussock Grasshopper** *Sigaus australis*. The former is patterned grey-green with pale stripes on its back. The latter is slightly larger, to 40 mm, and either green or brown and is widespread in alpine and subalpine habitats.

The **North Island Alpine Grasshopper** *Sigaus piliferus* is generally brownish-grey and up to 40 mm, it is widespread in mountainous areas in the NI. Found on both NI and SI is the **New Zealand Grasshopper** *Phaulacridium marginale*, this is much smaller, at 20 mm, than most other species and is found above 900 m. It has very variable colouration, mainly greens and browns and greys, depending on its surroundings.

KATYDIDS OR LONG-HORNED GRASSHOPPERS – TETTIGONIIDAE

Five species are found in New Zealand, most commonly the **Katydid** or **Kiki Pounamu** *Caedicia simplex*, which is widespread and probably native and is commonly found in gardens. It is normally light green and up to 45 mm long. Much smaller, at 15–20 mm, are the **Field Grasshoppers** *Conocephalus* spp., which have a very high-pitched 'song' and are either green or brown.

CRICKETS – GRYLLIDAE

Generally brown to near-black, largely nocturnal and only occasionally flying, there are eight species, all native. Most common is the **Black Field Cricket** or **Piharienga** *Teleogryllus commodus* which has a short 'chirp, chirp' song and is found from Kaikoura northwards. There are also four species of **Small Field Cricket** *Bobilla* spp., which are only 10–12 mm long and have a much lower frequency 'song' than most.

CICADAS – HEMIPTERA

There are some 40 species of Cicada (Cicadidae), all endemic. On hatching, the nymphs burrow into the ground, spending three or more years there. The Cicada 'song' is the loudest noise made by any insect anywhere and is made only by the males. Adults feed on plant sap. Males die soon after mating. Wings are always transparent. The commonest and most widespread is the **Chorus Cicada** or **Kihikihi Wawa** *Amphipsalta zealandica* which can be up to 25 mm long with a wingspan of 80 mm. The body is green with dark markings. Its song, which starts in January, is a continuous 'zeeep-zeeep-zeeep', often in chorus.

Rather smaller, and with olive body colouration, is the **Clapping Cicada** *Amphipsalta cingulata*. Its song is a long, multi-syllabic 'zeeeeep zap-zap-zap-zap'. It does not normally sing in chorus and is found only in the North Island. With the same range is the **Snoring Cicada** *Kikihia cutora*, which is bright green with yellow or orange markings and a distinctive 'snoring' song 'tea-tea-tea-tea tooo-tooo-tooo'.

The **Chirping Cicada** *Amphipsalta strepitans* is smaller, with reddish brown markings. It appears in November and its song is a repeated 'che-che-che'. While that of the **Little Grass Cicada** *Kikihia muta* is a repeated 'zit zit zit'. Most unusual of all is the small **High Alpine Cicada** *Maoricicada nigra* which is found above 1200 m virtually to snow line. Its song is a two-syllabic 'ur-zip ur-zip'.

Migratory Locust

Northern Tussock Grasshopper

New Zealand Grasshopper

Southern Tussock Grasshopper

Katydid

Field Grasshopper

Black Field Cricket

Chorus Cicada

Little Grass Cicada

BEES, WASPS AND ANTS – HYMENOPTERA

Hymenoptera refers to the translucent or membranous wings. These insects are mostly narrow-waisted.

BEES AND WASPS

There are 2000 to 3000 species of native wasps and bees in New Zealand, most of them endemic, varying in size from the 40 mm-long sawflies and wood wasps, to microscopic parasitic wasps less than 1 mm long. There are also many introduced species. Wasps are generally hairless while bees are nearly always hairy. Most wasps are parasitic, their larvae feeding on other invertebrates, bee larvae normally feed on pollen. Adults of both wasps and bees generally feed on nectar and similar sugary food.

NATIVE BEES

Native bees are widespread and found in all habitats.

The 30 or so species of native bee are solitary. There are 18 species of **Hairy Colletid Bees** *Leioproctus* spp. at 12 mm, slightly smaller than honey bees. All are black apart from *L. fulvescens* from the South Island, which has thick orange-yellow hair. They nest in the ground, making small tunnels (watch out for small soil piles). *L. metallicus* nests in sandy coastal areas.

The seven species of the genus *Hylaeus*, are smaller (9 mm) and are black with yellow markings on the head and thorax. They nest in holes in trees and branches and carry the pollen in their stomach.

Smaller still, at less than 8 mm long, are four species of the genus *Lassioglossum*, also ground-nesters, they are generally black or dark green and slightly hairy.

SOCIAL BEES – APIDAE

Bumble Bee *Bombus* spp. I
PI ROROHU

Four species, all introduced, originally to help pollinate clover. Readily distinguished by their size, shape, colour and by being noisy. The queen, at 25 mm, is significantly larger than the workers, hibernates during the winter. Can sting, but are not aggressive.

Honey Bee *Apis mellifera* I
PO HONI

A small (12 mm) light brown bee, first introduced in 1839 and found throughout the country, one of the few truly beneficial introduced species. Its sting can be painful.

SOCIAL WASPS – VESPIDAE

All 4 species of social wasp are introduced. They build 'paper' nests from wood and all can sting. Commonest are the **Common Wasp** *Vespula vulgaris* and the **German Wasp** *V. germanica*, with their distinctive yellow and black abdomens, the Common has alternate band while the German is similar but with additional black dots. Both feed on honeydew produced by the **Sooty Beech Scale** *Ultracoelostoma* spp., often in considerable numbers, preventing native insects, lizards and birds from using this important food resource. Two other social wasps, the **Asian Paper Wasp** *Polistes chinensis* and the **Australian Paper Wasp** *P. humilis* are also found mainly in the north. All are potential threats to native invertebrates.

PARASITIC WASPS – ICHNEUMONIDAE
NGARO WHIORE, ICHNEUMON WASP

Ichneumon flies or wasps are the largest group of wasps, often recognizable by the female's long or very long ovipositor. This is used to lay eggs in or on an invertebrate, on which the larva then feeds, often slowly so as not to kill its host too quickly. Most species are very host-specific and the larvae may feed on the eggs, larvae or pupae of the host species. Some species are very useful for pest control.

Hairy Colletid Bee

Bumble Bee

Honey Bee

Common Wasp

German Wasp

Ichneumon Wasp

Asian Paper Wasp

HUNTING WASPS – EUMENIDAE, POMPILIDAE, SPHECIDAE
NGARO WIWI

Some 30 species of hunting wasp are found in New Zealand. The female paralyses her prey with a sting, drags it to her nest and then lays her single egg on it.

Black Hunting Wasp *Priocnemis monachus* **N**
Widespread, especially December through February.

Medium-sized black wasp (BL 18–20 mm), nests in clay banks and preys mainly on tunnel web and trapdoor spiders. Adult feeds on fruit and nectar especially Manuka and Kanuka.

Mason Wasp *Pison spinolae* **N**
Widespread NI, SI and Stewart Is.

Small near-black wasp (BL 15 mm), female makes nest of mud, often on house exteriors or even indoors, preys mainly on orbweb spiders, and seals prey into individual nest. Adults feed on nectar.

SAWFLIES AND WOODWASPS – SYMPHYTA
Some 10 species, mostly introduced, many are large, to 40 mm, and very noisy. Grubs feed on wood and are themselves often preyed on by parasitic wasps. Often very large ovipositor, but they do not sting.

ANTS – FORMICIDAE
POPOKORUA, TOROTORO

There are 20,000 species of ant worldwide, but only 11 native species, along with 30 introduced species. The queens of all ants are significantly larger than the worker ants.

NATIVES
Most common is the **Southern Ant** *Monomorium antarcticum*, workers are 3–4 mm long, orange to black in colour and widespread in all habitats throughout the country.

Equally common but slightly larger is the **Striated Ant** *Huberia striata*, which varies from reddish-yellow to black but with variegated markings. Both species are found in association with sap-sucking insects that produce honeydew. The largest native ants are *Pachychondyla castanea* and *P. castaneicolor* whose workers are reddish brown and up to 6 mm long, they are more likely to be found in native bush.

INTRODUCED
Most are from Australia, many are pests and can cause serious ecological disruption. Commonest is the **White-footed Ant** *Technomyrmex albipes*, a small, 2.5–3 mm long, black or brownish-black ant with yellowish-white feet, and no sting. Also widespread is the **Argentine Ant** *Linepithema humile*, one of the 100 most invasive species in the world. It is slightly smaller than the White-footed Ant and forms very large colonies with many queens, they can displace native ants and other invertebrate species, they bite but do not sting.

In contrast, the **Red Imported Fire Ant** *Solenopsis invicta* is, at 1.2 mm, minute. It is equally invasive but has a very painful sting and large numbers of fire ant stings have been known to kill humans. Fire ants are voracious feeders on other invertebrates but have few natural predators and can be a serious environmental, agricultural and urban pest. They have been found several times in New Zealand, but have not yet become established. If you should find any, inform the Department of Conservation immediately.

DRAGONFLIES AND DAMSELFLIES -ODONATA

Dragonflies are generally more heavily built than Damselflies, they fly faster and more erratically, have eyes very close together and sit with outstretched wings while Damselflies always rest with them closed. The nymphs of all species develop in freshwater ponds and lakes where they are ferocious hunters. The adults feed on other flying insects, flies, bees and even cicadas.

DRAGONFLIES – ANISOPTERA
Found on all three main islands to heights of 1500 m.

There are 11 endemic species, plus some occasional summer visitors from Australia. The

Black Hunting Wasp

Mason Wasp

Red Percher Dragonfly

White-footed Ant

Bush Giant Dragonfly

largest native is the **Bush Giant Dragonfly** or **Kapokapowai** *Uropetala carovei* (see photo previous page), with a body length of 80–90 mm and a wingspan of up to 130 mm. It has a distinctive dark brown or black body with pale yellow markings on the thorax and at each joint in the abdomen.

Somewhat smaller is the native **Red Percher** *Diplacodes bipunctata* which is 30–35 mm long with a reddish body and tint to its wings. Found mainly in the North Island but as far south as Haast in the South Island. The **Baron Dragonfly** *Hemianax papuensis* is a relatively recent arrival; its body is 65–70 mm long, golden yellow and brown with very distinctive markings. This too is found mainly in the North Island.

DAMSELFLIES – ZYGOPTERA

Both species below are widespread by still water from the coast to alpine areas, NI and SI.
Six species are known in New Zealand. Rather more graceful than Dragonflies, Damselflies have a longer body in proportion to their wings. Often seen mating, where the male clasps the female behind the thorax. The **Blue Damselfly** or **Kekewai** *Austrolestes colensonis* is the largest species with a body length and wingspan of 45 mm. The males are blue-black and the females rather greener in contrast to the smaller **Red Damselfly** or **Kihitara** *Xanthocnemis zealandica* which has a body length of 30 mm and a wingspan of 40 mm.

OTHER INSECTS

APHIDS – APHIDIDAE

WEO, KUTIRIKI OR PEPERIKI
Aphids are very small soft-bodied insects that feed on plant sap. Some species have wings, others flightless or with winged stages only. Some 2500 species worldwide with 80 or so in New Zealand, mainly introduced. Native species tend to be quite plant specific.

STICK INSECTS – PHASMATODEA

RO, WHE
A wonderfully camouflaged and adapted insect of 9 genera with some 22 species, all endemic and all flightless. Males are very much smaller. Females are generally more numerous and reproduce by parthenogenesis, without mating. Indeed, with most species of the genus *Acanthoxyla* males have never been identified and they may reproduce entirely asexually. Stick insects eat plant leaves, feeding mainly at night, and when resting they bring front legs together with antennae to reduce chance of detection.

Giant Stick Insect   *Argosarchus horridus*   **EG**
Widespread NI and SI in lowland bush and also in parks and gardens.
The largest New Zealand species, with a very long grey-brown to greenish-brown rather spiny body (BL F to 145 mm/M to 100 mm). Feeds on leaves of Manuka and Kanuka, and also on Pohutukawa, Rimu, Totara and Rata, as well as introduced species. Can reproduce parthenogenetically, that is from unfertilized eggs, but produces only female offspring by this method.

Common Stick Insect   *Clitarchus hookeri*   **EG**
Widespread NI and SI, mainly lowland bush, parks and gardens.
Commonest stick insect (BL F 90–100 mm/M 60–65 mm) comes in two colour phases or variations, green or brown, feeds on Manuka, Kanuka and Pohutukawa.

Spiny Stick Insect   *Mimarchus* spp.   **E**
Otago northwards to Wairarapa.
Brown coloured with bumpy skin (BL F 55–60 mm/M 45–50 mm) feeds mainly on Bush Lawyer and sometimes roses.

Blue Damselfly   Red Damselfly

Giant Stick Insect

Spiny Stick Insect   Common Stick Insect

COCKROACHES – BLATTODEA

KOKOROIHE

A worldwide group of some 4000 species of which 40 or so are found in New Zealand. Most are native and found in bush, but you are likely to see introduced species. Some, such as the **Giant Cockroach** *Blaberus giganteus* and **American Cockroach** *Peroplaneta americana*, are tropical and so survive only indoors. The **Gisborne Cockroach** *Drymaplaneta semivitta*, recognizable by the cream-coloured lateral markings, is an Australian import. There are many native cockroaches, including the **Black Cockroach** or **Kekerengu** *Platyzosteria novaeseelandiae*, or the **Winged Bush Cockroach** *Parellipsidion latipennis* or one of the many **Flightless Bush Cockroaches** *Celatoblatta* spp. Most are found in native bush or forest, occasionally on the coast, but some species are found in alpine areas. None of the native species are pests.

PRAYING MANTIS – MANTODEA

RO, WHE

There are only two species of this remarkable insect in New Zealand, one endemic and one introduced. The forelegs, which are barbed, are used to hold prey, which includes flies, wasps, moths and other invertebrates, and on occasion the female will eat her mate, who is significantly smaller, during copulation.

New Zealand Praying Mantis *Orthodera novaezealandiae* E

Widespread in lowland bush, also parks and gardens.

Normally bright green in colour with distinctive blue dots on inside of forelegs (BL F 45 mm, M smaller). Found on top of leaves, praying for a meal.

African Praying Mantis *Miomantis caffra* I

Parks and gardens northern half of NI.

A recent (1978) introduction, larger than the endemic species (BL M 45 mm, F larger). Colouration pale green but sometimes brown. Lacks blue spots on forelegs.

New Zealand Glow Worm *Arachnocampa luminosa* E

TITIWAI, PURATOKE

Widespread throughout New Zealand, in caves and banks, perhaps the best known site is the Waitomo Caves in the NI.

Not actually a worm, but the larva of a fungus gnat or fly. The larvae live in a tube like 'nest' in a cave or bank, often shaded or partly concealed. They let down a number of sticky threads that entrap unwary insects attracted to the light emitted by the fly larva. The adult flies live for only two or three days. The female uses a light on its tail to attract a mate. The sight of a large number of these 'glow worms' can give the impression of a city of lights viewed from afar, or on the roof of a cave, like the Milky Way.

Sooty Beech Scale *Ultracoelostoma* spp. **EG**

Occurs throughout the country, especially on Black Beech *Nothofagus solandri*.

A sooty-mould fungus grows on beech bark thanks to the sweet, sticky secretion, called honeydew, produced by Sooty Beech Scale. This mould covers the scale insects so that they are invisible apart from the long filamentous anal tube which they use to excrete the honeydew. The females have four distinct life stages (known as instars): 1) 'crawlers' attach themselves to the trunk, bore into it and drink the sap, they excrete a waxy substance which hardens to form a protective shield; 2) and 3) they secrete the honeydew; 4) mating occurs, the female produces eggs within her body and then dies. The larvae hatch and exit the parent's body via the anus, and seek a new crack to live in. Males have five instars: 1) and 2) are as the female, but they emerge from the sheild as a 'prepupa' (3) and pupates (4) from whence they emerge with wings (5), at which stage they mate with females. Adult males have a deep reddish-pink body about 3–4 mm long with purplish pink wings with a span of 8 mm. Honeydew produced by the Sooty Beech Scale is an important contributor to the forest food supply; it has been estimated that a hectare of beech wood can produce between 3500 and 4500 kg dry weight of honeydew per year. This is a valuable food source for native fauna, including birds, during winter months when there are few, if any, flowers available. There are indications that introduced wasps can consume up to 90% of the available honeydew and prevent native birds from feeding. The actual impact of this is uncertain but likely to be deleterious.

Black Cockroach

Flightless Bush Cockroach

Gisborne Cockroach

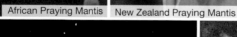

African Praying Mantis   New Zealand Praying Mantis

New Zealand Glow Worm

Sooty Beech Scale

THE SEASHORE

With over 15,000 km of coastline, and no place more than 140 km from the sea, there are plenty of opportunities to beachcomb or to wonder at the varied life in rockpools. The coasts of New Zealand are rich in animal and plant life, much of it endemic. This section is intended to give a taster of what is there.

MOLLUSCS

There are four common groups of molluscs, commonly referred to as 'shells': **Gastropods**, which includes most of the spiral shaped shells as well as limpets and abalones; **Bivalves**, whose shell is divided into two similarly sized, but different shaped halves, these include scallops, clams, mussels and oysters; **Chitons**, which look a little like an armour-plated slug; **Cephalopods**, which include squid, octopus and nautilus.

GASTROPODS

The **True Whelks** are molluscs with a tall conical structure, and a smooth-lipped or flanged opening. Whelks are predators and scavengers, feeding on other molluscs and able to drill through their shells. They are not large, but can be up to 50 mm long, with often quite knobbly shells.

Catseye *Turbo smaragdus* is a common, rather squat, snail-shaped mollusc which is clearly identified by the blue-green eye-like operculum that closes the opening. Of similar size are the **Grooved Topshell** *Melagraphia aethiops*, and the very dark **Black Nerita** *Nerita atramentosa*, both of which are spiralled vertically, while the Catseye is spiralled laterally. Much smaller are the periwinkles. Two common ones are the **Blue-banded Periwinkle** *Austrolittorina antipoda* and the **Grooved Periwinkle** *A. cincta*, both are found on rocks close to the high-tide mark, often in quite large numbers.

Limpets are simple pyramid-shaped shells which are found on inter-tidal rocks. The **Ornate Limpet** *Cellana ornata* is found in the mid- to high-tide zone and has a distinctly ribbed shell, while the **Radiate Limpet** *C. radians* is found lower down and is smoother, often with yellow or green colouration. Limpets feed on algae and seaweeds and hold very tightly to rocks, enabling them to withstand strong wave action.

BIVALVES

Most bivalves are found either firmly attached to rocks, or buried deep in sand and mud. On the rocks are two common species of mussel; the very small **Little Black** or **Flea Mussel** *Xenostrobus pulex*, which is found in profusion on rocks almost to the high-tide mark, and the much larger and regularly harvested **Green-lipped Mussel** *Perna canaliculus*, which is found around the low-tide mark. Also widespread is the **Pacific Oyster** *Crassostrea gigas*, which clings so closely to the rocks that it almost seems part of them. The glue that these molluscs produce to anchor themselves to the rocks is superior to any synthetic glue.

Many of the bivalves that live in the sand or mud are much prized for food. Among these, in the mid-tide sand, are **Toheroa** *Paphies ventricosa* (the fishing of which is currently prohibited) which can be up to 120 mm long. Further down the beach are **Tuatua** *P. subtriangulatum* and, in the estuaries, you will often see people fishing for shellfish, especially the common bivalves **Pipi** *P. australis* and the **Tuangi** or **Common Cockle** *Austrovenus stutchburyi*.

Speckled Whelk

Catseye

Black Nerita

Grooved Periwinkle

Radiate Limpet

Little Black Mussel

Ornate Limpet

Green-lipped Mussel

Tuatua

Pacific Oyster

Tuangi

MOLLUSCS   125

BEACH SHELLS

Two bivalves feature among the many mollusc remains washed up on the beach. One is the attractive **New Zealand Scallop** *Pecten novaezealandiae*, with one flat and one concave half to their shells. They live offshore and are fished commercially. Also common on some beaches are the very attractive triangular **Horse Mussel** *Atrina zelandica*, a type of pen shell which can be up to 50 cm long. Among gastropods, watch out for whole or part shells of the **Blackfoot Paua** *Haliotis iris*, a much sought after food, with its attractive mother-of-pearl lining that is popular with craft workers and jewellers. **Cook's Turban** *Cookia sulcata* is a large, broadly conical mollusc with an attractive mother-of-pearl sheen on the outside, but which is normally covered by barnacles and other marine growth. Watch out also for **Turret** or **Tower shells** Turitellidae, which are long and spirally pointed, the larger **Volutes** *Alcithoe* spp. and the rather knobbly **Tritons** *Charonia* spp.

CHITONS

These multivalve molluscs are stuck firmly onto the rocks well above the mid-tide line and their segmented, armoured shells make them look rather like tiny legless armadillos. There are a number of species. Possibly the most widespread is the attractive **Snakeskin Chiton** *Sypharochiton pelliserpentis*.

CRUSTACEANS

BARNACLES

Barnacles are small crustaceans that anchor themselves to rocks with a protein-based glue (that we have been unable to emulate). There are four common species, of which three have typical off-white conical shells: **Small Barnacle** *Elminius modestus*, which is just that and somewhat star-shaped; **Columnar Barnacle** *Chamaesipho columna*, which has rather regular, near-vertical sides; **Plicate Barnacle** *Epopella plicata* is similar but with grooved sides; and the **Stalked** or **Goose Barnacle** *Lepas anatifera*, which is black and white and has a long flexible stalk.

CRABS

The **Purple Shore Crab** *Leptograpsus variegatus* is one of the most widespread crabs. It is very handsomely coloured, as is the rather more heavily built **Marbled Rock Crab** *Hemigrapsus edwardsi*. Smaller and rather muddy-looking is the **Tunnelling Mud Crab** *Helice crassa*, whose eyes stick up vertically when not in its burrow.

If you see a whelk, triton or turban shell moving quite quickly it is almost certain that a **hermit crab** has taken up residence in the shell. There are several species; the commonest is *Pagurus novizelandiae* which can be found in a variety of 'houses'.

Purple Shore Crab

Marbled Rock Crab

Tunnelling Mud Crab

Hermit Crab

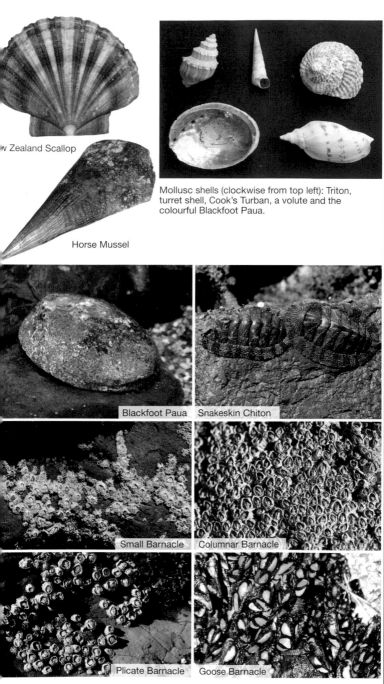

New Zealand Scallop

Horse Mussel

Mollusc shells (clockwise from top left): Triton, turret shell, Cook's Turban, a volute and the colourful Blackfoot Paua.

Blackfoot Paua

Snakeskin Chiton

Small Barnacle

Columnar Barnacle

Plicate Barnacle

Goose Barnacle

ECHINODERMS

This group includes starfish, brittle stars, sand dollars and sea cucumbers. Starfish are quite common, especially the brightly coloured **New Zealand Cushion Star** *Patiriella regularis*, generally well hidden in a rock crevice, also the **Reef Star** *Stichaster australis*, which has nine arms. Some species have up to 11 arms. Closely related, but with very long, thin arms, are the brittle stars, a quite common one is the **Mottled Brittlestar** *Ophionereis maculata*.

While beachcombing you will often come across remains of two common echinoderms. These are the **Kina** or **New Zealand Sea Urchin** *Evechinus chloroticus*, a rock dweller, and its sand-dwelling close relative the **Snapper Biscuit** or **Sand Dollar** *Fellaster zelandiae*, which lives in deeper water. What you are seeing is just the 'test' or skeleton, which is covered with a soft membrane when the animal is alive.

INVERTEBRATES

To find invertebrates, try looking under seaweed and driftwood. You will often find a number of small arthropods, the commonest of which is the **Sandhopper** *Talorchestia quoyana,* a small, transparent animal that hops, and the dark brown **Seashore Earwig** *Anisolabis littorea*, which is larger and darker than the more common, land-based introduced European earwig.

SEAWEEDS

New Zealand boasts several hundred species of seaweed – green, brown, red and yellow, Two common seaweeds found in tidepools are the rather bushy *Halopteris virgata* and *Codium fragile,* with its unusual round, green, stick-like shoots. Keep an eye open also for the attractive *Xiphophora gladiata*, with its flattened dividing branches and for the all-too-familiar **Sea Lettuce** *Ulva lactuca*, which when present in quantity is generally an indication of pollution, agricultural or domestic.

After a storm, large amounts of seaweed may be washed up on the beach, including the impressive **Bull Kelp** *Durvillaea antarctica*. It can have large, leathery fronds more than 10 m long, with an amazing honeycomb structure giving it impressive buoyancy. Another very recognizable seaweed is **Neptune's Necklace** *Hormosira banksii*, looking like a bead necklace, while *Cystophora scalaris* is unusual with its zigzag stem. If you find an old mussel shell with a bushy growth, it is not a seaweed but **Mussel's Beard** *Amphisbetia bispinosa*, an hydroid, a relative of corals.

TIDEPOOLS

Tidepools are often rich in marine life including, crabs, starfish, sea urchins, seaweeds and small fish. One widespread fish is the **Mottled Triplefin** or **Twister** *Bellapiscis lesleyae*, though they are often very well camouflaged. Sea anemones are common, the most easily recognizable is the **Red Anemone** *Isactinia tenebrosa*.

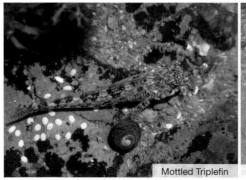
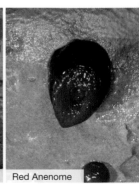

Mottled Triplefin   Red Anenome

New Zealand Cushion Star   Reef Star

Mottled Brittlestar   Kina   Sand Dollar

Sandhopper   Seashore Earwig   *Codium fragile*

Xiphophora gladiata   Sea Lettuce

Cystophora scalaris   Bull Kelp

Mussel's Beard   Neptune's Necklace

NATIVE TREES AND SHRUBS

While settlers have destroyed three-quarters of New Zealand's native bush, no species of tree or shrub has become extinct. We are therefore still able to enjoy the large variety of species that make up the very varied flora of the native bush. New Zealand has more than 260 species or subspecies of native tree and a further 640 native shrubs, some 80% of these are endemic. (The symbol **E** after the botanical name indicates it is an endemic species and **EG** that it is an endemic genus. **N** = native species.) Compare this to 33 native tree species in the British Isles.

IDENTIFICATION

Identifying New Zealand trees and shrubs is not easy: first, there are a lot of them; second, most of them are evergreen; third, many of the larger ones disappear through the canopy leaving only the trunk visible; and, finally, many species have juvenile forms quite unlike the mature plant. In order to help with identification, tree and shrub species are grouped mainly by visual characteristics, especially leaf shape and form, in the following way.

SMALL-LEAVED LARGE TREES P 132
- Large or very large trees with small, thin leaves — native conifers
- Large trees with small rounded leaves — native beeches.

TUFTED OR PALM-LIKE LEAVES P 140

PALMATE OR FINGER-LIKE LEAVES P 142

LONG NARROW LEAVES P 144

TOOTHED LEAVES P 150

PINNATE OR SIMILARLY PAIRED LEAVES P 156

GLOSSY, FLESHY OR LEATHERY LEAVES P 158

VARIED OR COLOURED LEAVES P 164
Leaves that are not green or have more than one colour.

SMALL LEAVES P 166

HEBES P 172
A large and distinctive family of small to medium-sized shrubs.

TREE DAISIES P 174
Distinctive yellow- or white-flowered shrubs with a daisy flower.

ALPINE AND SUBALPINE SPECIES P 178
Found almost exclusively in high alpine or subalpine regions.

Each species within a category is then listed in approximate order of maximum height, i.e., larger trees first, low shrubs last.

Where a species could fit into two or more categories, it will be listed in all sections but only described in one. For instance, Whauwhaupaku or Five-finger *Pseudopanax arboreus* will be described under Palmate or Finger-like Leaves, but because its leaves are also toothed, it will be cross-referenced under Toothed Leaves also.

Te Matua Ngahere, 'Father of the Forest', the second largest remaining Kauri, with a trunk diameter of over 5m. Waipoua Forest, Northland.

SMALL-LEAVED LARGE TREES
NATIVE CONIFERS

Conifers are trees that do not have flowers but reproua by producing pollen cones and seed cones. The pollen fertilises the seed cone where the seeds develop. There are 20 species of native conifer, divided into five families. Many, though by no means all, are very tall with impressively straight trunks and could easily be considered the 'Kings' of the forest. H = height / TD = trunk diameter

ARAUCARIACEAE
Kauri *Agathis australis* E

Northland, Coromandel and Kaimai Mamaku: esp. Waipoua Forest, Trounson Kauri Park, Warkworth, Ngaiotonga Scenic Reserve and Russell Forest, Waiau Falls Scenic Reserve.

An immense tree, without parallel (H to 75 m/TD to 7 m). The second largest tree in the world by trunk volume. Massive, impressively straight trunk branching out 20–30 m above the ground into an immense crown of branches which become festooned with epiphytes. Leaves of mature Kauri are almost rectangular, 20–35 mm long by 15–20 mm wide. The seed cones are spherical, up to 80 mm in diameter, while the pollen cones are stubby and up to 4 cm long. The bark is very distinctive, with flakes falling off creating a colourful embossed surface which helps to prevent vines and epiphytic plants from attaching themselves to the trunk.

Young Kauri are known as 'rickers' and are narrowly conical, grow at about 30 cm a year and start to mature at 100–125 years of age. Their leaves are much longer, 50–100 mm, and narrower, 5–10 mm, than mature trees, and are often slightly copper-coloured.

The two largest remaining Kauri are thought to be about 2000 years old, they are: **Tane Mahuta 'Lord of the Forest'** Trunk girth: 13.77 m / Trunk height: 17.68 m / Total height: 51.2 m / Trunk volume 244.5 m³; and **Te Matua Ngahere 'Father of the Forest'** Trunk girth: 16.41 m / Trunk height: 10.21 m / Total height: 29.9 m / Trunk volume: 208.1 m³

According to Maori mythology, Tanemahuta was the oldest son of Ranginui the Sky Father and Papatuanuku the Earth Mother. Tane forced them apart, allowing light into the world, and indeed these massive trees do look as though they are supporting the sky.

NATIVE CEDARS – CUPRESSACEAE
Kawaka *Libocedrus plumosa* E
KAIKAWAKA

Lowland forest, north of Bay of Plenty, and in the northwest of SI.

Tall conical tree (H 25 m/TD 1.2 m). Young trees are branched to the ground and have a 'weeping' appearance. Mature trees have bare trunks for at least half of their height. Bark is dark reddish-brown and falls off in long vertical strips. The leaves, which are actually flattened branchlets, are pinnate and much wider on juvenile trees (7 mm) compared to mature ones (3 mm). Pollen cones, up to 3–5 mm long, and seed cones, up to 4–6 mm long, form on the tips of the branchlets.

Pahautea *Libocedrus bidwillii* E
KAIKAWAKA, MOUNTAIN CEDAR

Wetter montane forest from Bay of Plenty southwards, commoner on the west coast in SI.

Similar to but smaller (H 20 m/TD 1 m) than *L. plumosa*; mature trees often have a slight 'broccoli' appearance. The branchlets of juveniles are flattened but only 3 mm wide while in mature trees they are cylindrical. The bark is very similar to, and both cones are larger than, *L. plumosa*. Often forms buttress roots.

MIRO AND MATAI – PRUMNOPITYACEAE
Miro *Prumnopitys ferruginea* E
BROWN PINE, TOROMIRO

Widespread, lowland and lower montane forest on all main islands, especially west coast NI and SI, and Stewart Is.

Tall, upstanding, round-headed forest tree (H 30 m/TD 1 m). Bark is grey and scaly, but brown when scales fall off. Leaves thin, 25 mm by 3 mm, pointed, slightly curved and flattened into two rows. Juvenile leaves are more pointed, slightly longer and almost fern-like. Pollen cones, to 15 mm, stand erect from branchlets; seed cones are on short stems producing large fleshy red seeds, a favourite food for Kereru.

Kauri bark [leaves and cones]

Kauri

Miro fruit

Kawaka

Miro

Pahautea [cones]

Matai *Prumnopitys taxifolia* E
BLACK PINE

Widespread in lowland forest especially on alluvial and volcanic soils, NI and SI only.
Very large, round-topped forest tree (H 25 m/TD 2 m) with a distinctive brown bark which falls off in somewhat rounded flakes leaving bright red-brown patches. The leaves are similar to *P. ferruginea*, but are straight, thinner, rounded and whitish on the underside. Juvenile plants in the understorey may be divaricate with leaves often brown or reduced to scales. Pollen cones are yellow and in groups on spikes. Seed cones are also grouped on spikes and are purple-black when mature.

NATIVE PODOCARPS – PODOCARPACEAE
Totara *Podocarpus totara* var. *totara* E
Widespread in lowland and montane forest in NI and SI, particularly alluvial soils.

Very large (H 30 m/TD 2 m), densely foliaged tree characteristic of lowland forest, and often left as remnant tree in agricultural land. Characteristic fibrous, heavily furrowed and ribbed bark. Leaves small (30 mm by 4 mm) and pointed, similar in appearance to European Yew *Taxus baccata*. Pollen cones up to 20 mm are individual or in small clusters and seed cones develop a fleshy base, maturing through red to black. The wood is prized by Maori and Pakeha alike – it is strong, durable and easily worked.

Westland Totara *P.t.* var. *waihoensis* is found only on the west coast of SI from Waihoa River south to the Cascade Range.

Needle-leaved Totara *P. acutifolius*, a low-growing shrub with similar but narrower leaves to *P. totara* is found in the northern end of SI, particularly at higher altitudes and in beech forest.

Mountain Totara *Podocarpus cunninghamii* E
HALL'S TOTARA, THIN-BARKED TOTARA, *P. HALLII*

Found on all main islands, widespread lowland to subalpine forest, replaces *P. totara* at higher altitudes and on poorer soils.

Very similar to, but generally smaller than (H 20 m/TD 1.5 m) *P. totara* in general appearance, however its bark is thin and papery and peels off in long narrow strips. Adult leaves are 20–30 mm long and 3–4 mm wide, rather pointed and slightly prickly to the touch. Juvenile leaves are significantly longer, to 50 mm, and broader. Pollen cones to 25 mm, seed cones very similar to *P. totara*.

Manoao *Manoao colensoi* EG
SILVER PINE

High rainfall lowland and wetland areas in NI and in Westland.

Small to medium-sized tree (H 15 m/TD 60 cm) with grey bark which falls off in large flakes. Juvenile has long (6–12 mm) narrow leaves, while adult has yellowish-green leaves 1–2.5 mm long which are appressed to the twig. Pollen cones are reddish green on twig tips; seed cones are on different trees and are dark purple with a greenish collar.

Yellow Silver Pine *Lepidothamnus intermedius* E
Lowland and montane forest, especially poor soils or boggy areas. Scarce on NI but common on the west coast of the SI and Stewart Is.

A medium-sized (H 15 m/TD 60 cm) rather spreading tree, the bark is grey-brown and falls of in scales, reminiscent of Kauri but much smaller, leaving a patterned pitted surface. Leaves of juvenile are long, to 15 mm, narrow, and pointed but changing into short (3 mm) scale leaves in the adults. Pollen cones are reddish green and form on the tips of the branchlets. Seed cones are on separate trees and produce black seeds with a fleshy yellow collar.

Matai

Manoao [cones]

Mountain Totara

...otara [foliage, top, and bark]

Yellow Silver Pine, adult foliage

Pink Pine *Halocarpus biformis* **EG**

Higher montane and subalpine forest and scrub, especially boggy areas, from Coromandel south.

A small, often rather shrubby tree (H 10 m/TD 60 cm), the grey bark forms small thick flakes. Lower branches may root. Juvenile leaves are typically narrow (20 mm x 3 mm) and pointed while mature leaves are scale leaves about 2 mm in length, giving a bumpy appearance. Pollen cones are reddish green and form on tips of branchlets. Seed cones are on separate plants; seeds have fleshy yellow collar. Common name comes from the colour of the wood.

Rimu *Dacrydium cupressinum* **E**

RED PINE

Widespread in lowland and montane forest on all main islands.

Distinctive very tall (H 60 m/TD 1.5 m) straight trunked, somewhat conical, emergent tree with very characteristic drooping foliage. Bark flakes off in quite large pieces leaving distinctive contour-like pattern. Juvenile leaves are thin and up to 7 mm long on long, drooping twigs becoming short, 2–3 mm, scale leaves in mature trees. Juvenile trees very attractive with their 'weeping' appearance. Pollen cones are at the ends of twigs. Seed cones are on separate trees and have a fleshy red collar when ripe. Previously an important timber tree, lives to 1000 years or more.

Kahikatea *Dacrycarpus dacrydioides* **E**

WHITE PINE

Widespread in lowland, especially swampy, forests on all three main islands.

Tallest New Zealand tree. Very tall (H 60 m/TD 1.5 m) and graceful trunk often fluted with a twist, and with buttresses. Younger trees quite conical, older ones have rounded crowns. Bark grey and falls off in rounded flakes. Juvenile leaves to 7 mm long, flattened in two rows, mature trees have small scale leaves 1–2 mm long. Yellow pollen cones appear at ends of twigs. Seed cones are on separate trees producing smooth, round, red berry, enclosing a shiny black seed.

CELERY PINES – PHYLLOCLADACEAE

Celery Pines are so called because of their unusually shaped leaves which are in fact not leaves but flattened branchlets or phylloclades – leaf-stems.

Tanekaha *Phyllocladus trichomanoides* **E**

CELERY PINE, CELERY-TOPPED PINE

Lowland forest in NI, south to Hawke's Bay and Taranaki and northern SI.

A tall (H 20 m/TD 1 m) pyramid-shaped tree with distinctive celery-shaped rather leathery leaves which are flattened branchlets or phylloclades. Bark is smooth – grey when young, brown when mature and often encrusted with lichens. Pollen cones form in clusters at the tips of the branches and are red at maturity; the seed cones are small and form on the margins of the phylloclades and bear a single seed.

Toatoa *Phyllocladus toatoa* **E**

Lowland and montane forest south to Rotorua.

Medium-sized tree (H 15 m/TD 60 cm) similar in appearance to Tanekaha, but phylloclades are thicker and more shield-shaped with indentations along the margins. Trunk grey and slightly warty, with irregular horizontal ridges. Pollen cones in small clusters at branch tips, yellow when mature. Seed cones found at base of phylloclades.

Mountain Toatoa *Phyllocladus alpinus* **E**

MOUNTAIN CELERY PINE

Widespread, montane and subalpine forest, 900–1600 m in NI, lowland forest in south of SI.

Small, rather bushy tree (H 9 m/TD 30 cm) that may spread by lower branches taking root. Phylloclades are individual and stalked, somewhat diamond-shaped with irregular margins. Trunk grey with wrinkled appearance. Pollen cones smaller than *P. toatoa* and crimson at maturity. Seed cones develop on the phylloclade stalks.

Rimu

Pink Pine [foliage]

Rimu trunk

Rimu juvenile foliage

Kahikatea [male cones]

nekaha seed cones

Tanekaha pollen cones

Mountain Toatoa seed cone

Toatoa

SMALL-LEAVED LARGE TREES   **137**

NATIVE BEECHES – NOTHOFAGACEAE

There are four species of native beech, and two subspecies. All are evergreen and are related to the northern hemisphere beeches and birches, but do not visually resemble them. Beech forests often have a canopy formed of a single species with no subcanopy. This results in a very open forest floor with just a few small shrubs, along with ferns, mosses, fungi and epiphytes. Beech forests are the predominant forest on the South Island and generally flourish in cooler conditions, at higher altitudes and on poorer soils. Beech trees do not flower every year and they spread very slowly.

Hard Beech *Nothofagus truncata* E
TAWHAI RAUNUI, TAWAI RAUNUI
Lowland and lower montane forest from Northland south to northern Westland, excluding Taranaki.
The least common of the beeches. A large forest tree (H 30 m/TD 2 m) with buttress roots and rather rough and furrowed bark, sometimes covered in black mould that forms around scale insects which produce honeydew. The leaves are glossy green on top, rounded-elliptical and toothed, up to 35 mm long. Male flowers are red and appear in groups of one to three, female flowers yellowish-green. The wood is noticeably hard due to its high silica content.

Red Beech *Nothofagus fusca* E
TAWHAI RAUNUI, TAWAI RAUNUI
Lowland and montane forest from Bay of Plenty southwards, higher altitudes in the NI, lower in the SI.
A tall straight-trunked forest tree (H 30 m/TD 2 m) often with large buttresses. The bark is dark brown and rough. Leaves 40 mm, rounded-elliptical and toothed, green but turning red before falling off. Male flowers in groups of 1 to 8, long yellow stamens. Female flowers inconspicuous groups up to five. The wood is red when first cut.

Silver Beech *Nothofagus menziesii* E
TAWHAI, TAWAI
Lowland and montane forest from Bay of Plenty southwards, to treeline in wetter areas.
Very large forest tree (H 30 m/TD 2 m), occasionally buttressed, with a large spreading crown. The bark is smooth and silver-white in young trees but more grey, shaggy and often covered with lichens and moss in mature trees. Leaves are very small, 20 mm x 15 mm, shield-shaped and toothed. Male and female flowers are greenish yellow, the female being closer to the ends of the branches than the male. At higher altitudes it has a very stunted form. Host of the parasitic fungus *Cyttaria gunneri*.

Black Beech *Nothofagus solandri* var. *solandri* E
TAWAI RAURIKI
Lowland and montane forest from Central Plateau south to Fiordland.
Large forest tree (H 25 m/TD 1 m+), often in large exclusive stands, with rough grey bark, frequently covered in black mould resulting from the honeydew of aphids that feed on the sap. Leaves are small, to 40 mm, dark green above, paler below, and ovoid with blunt tips. Flowers are small and inconspicuous, female ones closer to branch ends than male ones, which may have red anthers.

Mountain Beech *Nothofagus solandri* var. *cliffortioides* E
TAWAI RAURIKI
Montane forest in NI, montane and lowland forest in SI, to the treeline, often in drier and poorer soils.
Similar but much smaller that *N.s.* var. *solandri* (H 15–30 m/TD 60 cm), with smoother bark becoming rougher with age. Sometimes covered in black mould from aphid honeydew. Leaves smaller and pointed and with pale hairs on underside. Male flowers may have red anthers and appear further from branch ends than females. Larger specimens are found in far south, often festooned with moss, both cushion type and long streamers of *Weymouthia*.

Red Beech flowers

Silver Beech foliage and flowers

Hard Beech [foliage]

Silver Beech

Black Beech [flowers]

Mountain Beech [foliage]

SMALL-LEAVED LARGE TREES **139**

TREES AND SHRUBS WITH TUFTED OR PALM-LIKE LEAVES

Nikau Palm *Rhopalostylis sapida* E
FAMILY ARECACEAE
Widespread in coastal lowland and lower montane forest south to Banks Peninsula on the east coast of SI and Greymouth on west coast of SI.
Tall, slender-trunked palm (H 10 m, rarely 15 m/TD 25 cm) with fronds or leaves extending upwards at 45 degrees or more, from a smooth vase-shaped structure formed by the leaf bases. Trunk is smooth but ringed and sometimes covered in epiphytes or vines. Leaves are up to 3 m long and more than 1 m wide, with individual leaflets up to 1 m long. Flowers appear on large inflorescences at base of fronds, pinkish in colour and turning into bright red berries.

Cabbage Tree *Cordyline australis* E
TI KOUKA
FAMILY LAXMANNIACEAE
Widespread in lowland forest margins, swamps and open country on all three main islands. Probably the most recognisable plant in New Zealand.
Unmistakeable, tufted ('Dr. Seuss') tree (H 12 m/TD 1.5 m). Mature trees have many-branched trunk with rough grey, cork-like bark. Leaves long, to 1 m, and narrow, occurring in tufts at the ends of branches. Flowers white and profuse on large bunched inflorescences producing pale purplish berries.

Forest Cabbage Tree *Cordyline banksii* E
TI NGAHERE
FAMILY LAXMANNIACEAE
Widespread in NI in lowland forest margins, banks and cliffs, also northwest of SI.
A shrub or short palm-like tree (H 4 m/TD 15 cm) with a large rosette of narrow, pointed leaves up to 2 m long, generally drooping as they mature. Leaves leave scars on trunk as they die and fall away. Small, fragrant, white flowers on large inflorescences up to 2 m long in late spring and summer. Berries white or bluish, 4 mm diameter.

Mountain Cabbage Tree *Cordyline indivisa* E
TOI, BROAD-LEAVED CABBAGE TREE
FAMILY LAXMANNIACEAE
Margins of montane forest and shrublands, from Auckland southwards apart from east and extreme south of SI.
Large palm-like plant (H 8 m/TD 20 cm), unbranched, with very large rosette of pointed leaves up to 2 m long and 15 cm wide. Dead fronds hang down in large skirts obscuring much of the trunk. Massive and distinctive inflorescence with finger-like branches covered with small cream or purplish flowers attached to a central stem that may be up to 1.5 m long.

Lowland Cabbage Tree *C. pumilio* is much smaller, has narrower rather yellowy leaves, grows to a maximum of 2 m and is found on lowland forest margins from Bay of Plenty northwards.

Mountain Neinei *Dracophyllum traversii* E
FAMILY ERICACEAE
Montane forest from Auckland south to northwest of SI.
A low shrubby tree (H 10 m/TD 60 cm) with distinctive red-brown papery bark, constantly peeling. Leaves, long, to 600 mm, narrow and downcurved are in candelabra-like tufts at the ends of the branches often with accumulated dead leaves hanging underneath. Small pinkish flowers on long, to 300 mm, terminal inflorescence.

Neinei *D. latifolium* is slender and less shrub-like with fewer but larger leaf tufts, and is found only in the northern half of NI.

Dracophyllum strictum is a small shrub to 2 m with leaves to 100 mm and white terminal inflorescences. Range similar to Mountain Neinei in lowland and montane forest.

See also Alpines for **Spreading Grass Tree** *Dracophyllum menziesii*, **Prostrate Grass Tree** *D. prostratum*, *D. pronum* and *D. recurvum*.

Nikau Palm   Cabbage Tree

Forest Cabbage Tree   Mountain Cabbage Tree

Mountain Neinei   Neinei   *Dracophyllum strictum*

TREES AND SHRUBS WITH TUFTED OR PALM-LIKE LEAVES   **141**

TREES AND SHRUBS WITH PALMATE OR FINGER-LIKE LEAVES

Puriri *Vitex lucens* E
FAMILY VERBENACEAE
Coastal and lowland forest south to Taranaki and Hawke's Bay.
Large, spreading forest tree (H 20 m/TD 1.5 m), often of great age with several trunks, often hollow. Bark is quite smooth and light grey-brown, but ribbed and furrowed with age. Leaves mainly five fingered with central one longer and stalked. Leaflets glossy and slightly wavy. Flowers are pink with long white stamens, berries large, to 20 mm, and red-orange, favoured food of Kereru. A member of the teak wood family.

Kohekohe *Dysoxylum spectabile* E
FAMILY MELIACEAE
Coastal and lowland forest especial further north, becoming coastal further south Marlborough Sounds only on SI.
A medium-sized under-canopy forest tree (H 15 m/TD 1 m). Bark quite pale and smooth, often partly covered in lichens. Leaves are large, to 400 mm, glossy green, seven fingered, leaflets slightly wavy on short stalks. Flowers on large hanging inflorescences, up to 300 mm, growing directly on the trunk and branches. Fruit are large, to 25 mm, round and green. They and take up to 12 months to mature and contain 3 orange-red seeds.

Raukawa *Raukaua edgerleyi* EG
FAMILY ARALIACEAE
Widespread in moist lowland forest on the three main islands.
Small understorey tree or shrub (H 10 m) often starting as an epiphyte. The leaves change as the tree matures; juvenile leaves are strongly aromatic, have 3 to 5 leaflets, the middle one often significantly the longest, deeply lobed, almost pinnate, on stalks up to 80 mm long. Adult leaves are very shiny, simple, up to 150 mm long on stalks up to 50 mm long and smooth margined. Scented when crushed. Flowers small, yellow in sprays at ends of twigs, berries round and dark purple.

Raukaua simplex EG
FAMILY ARALIACEAE
Widespread in moist forests from Firth or Thames southwards, including Stewart Is, at higher altitudes in the NI, lower in the SI.
Small many-branched forest tree or shrub (H 8 m). Juvenile leaves have 3 to 5 leaflets on stalks up to 100 mm long. Two subspecies; *R.s. simplex* has deeply lobed juvenile leaves becoming toothed in maturity; while *R.s. sinclairii* is multi-fingered and toothed as a juvenile becoming simple and toothed in maturity.

Pate *Schefflera digitata* E
PATETE
FAMILY ARALIACEAE
An understorey tree in lowland and montane forest on the three main islands, especially moist and sheltered places.
Small soft-wood tree (H 8 m) with 5–9 fingered leaves on long, to 250 mm, stalks. Leaflets, on stalks to 30 mm, are toothed, pointed and up to 200 mm long. Flowers small, greenish-yellow, on long drooping finger-like inflorescences, flowers late summer, early autumn. Berries round, white to purple.

Five-finger *Pseudopanax arboreus* E
WHAUWHAUPAKU, PUAHOU
FAMILY ARALIACEAE
Lowland and montane forest on the three main islands, especially margins and regenerating bush.
Small rather bushy tree (H 8 m) with distinctive 5- to 7-fingered, dark green slightly leathery leaves with stalks up to 200 mm long. Leaflets are up to 200 mm x 70 mm on stalks up to 50 mm and coarsely toothed. Bunches of small greenish flowers with dark purple buds, strongly scented, on branch tips June–Sept, turning into purple-black seeds in summer. Occasionally epiphytic on tree ferns.

Kohekohe

Puriri [flowers and fruit]

Raukawa

Raukaua simplex

Pate [fruit]

Five-Finger [flowers]

Five-Finger

TREES AND SHRUBS WITH PALMATE OR FINGER-LIKE LEAVES   **143**

Mountain Five Finger or Orihou *P. colensoi*, is very similar, however normally only 3–5 leaflets which have no stalks. Flowers pale yellow and in smaller clusters appear in spring.

Houpara *P. lessonii* is a small shrubby tree (H 6 m) found in coastal forest and shrubland from Gisborne northwards, it has leathery 3-fingered leaves on stalks up to 150 mm long. The leaflets are notched around the tip.

P. discolor is a small shrub (H 5 m) found in lowland forest and shrubland from Auckland northwards. It has 3–5 leaflets on stalks up to 80 mm. Leaflets are unstalked and coarsely toothed, often reddish-green in colour.

P. laetus is a small shrub or tree with large 5- to 7-fingered, slightly purple leaves on long stalks to 200 mm. Leaflets are up to 250 mm long on 30 mm stalks. Found in lowland to montane forest from Coromandel to Taranaki.

Wharangi *Melicope ternata* E
FAMILY RUTACEAE
Coastal and lowland forest margins, restricted to northeast of SI.
A small tree or shrub (H 6 m) with alternate, stalked, yellowish green, 3-fingered leaves, the leaflets are also stalked. The flowers are pale yellow-green with 4 petals and in clusters at the branch tips. Forms dry seed capsules with 4 pods each with a shiny black seed.

TREES AND SHRUBS WITH LONG NARROW LEAVES

Takes precedence over toothed, pinnate, leathery.

Tawa *Beilschmiedia tawa* E
FAMILY LAURACEAE
Lowland and montane forest, south to northern half of SI. Dominant tree in much of NI.
Large dominant canopy tree (H 30 m/TD 1.2 m) with smooth dark grey bark. Leaves opposite, narrow lanceolate, whitish below. Young leaves reddish. Flowers very small, greenish-white on panicles near branch ends. Fruit large 25 mm, purple, favoured food of Kereru and Kaka.

Rewarewa *Knightia excelsa* EG
NEW ZEALAND HONEYSUCKLE
FAMILY PROTEACEAE
Widespread lowland and montane forests in NI and in Marlborough Sounds in SI.
Tall canopy tree (H 30 m /TD 1 m). Younger trees are almost conical while mature trees have bushy crown and tall branchless trunk. Bark grey and smooth. Leaves leathery, narrow, pointed, elliptic, with a serrated edge. Flowers small and red, with cream-tipped pistil crowded onto a 10 cm long raceme giving bottle brush appearance. Fruit winged held in 40 mm long pod which may take up to a year to open.

White Maire *Nestegis lanceolata* E
FAMILY OLEACEAE
Lowland and montane forests to 600 m south to Nelson and Marlborough.
Small to medium-sized canopy tree (H 15 m/TD 1 m) with greyish vertically furrowed bark. Leaves long, to 120 mm, narrow, pointed, midrib flush on top, prominent on underside. Flowers yellowish on short racemes from branches or leaf axils. Fruit reddish drupes. Immature leaves long and very narrow, to 4 mm.

Black Maire *N. cunninghamii* is similar but larger (H 20 m), trunk cracked and fissured, leaves longer and wider, flowers hairy, drupes sometime yellow. Lowland forests only.

Mountain Maire *N. montana* (also Oro-oro or Narrow-leaved Maire), is much smaller, (H 10 m) bark rougher and more furrowed, leaves very narrow, flowers and fruit smaller.

Pseudopanax discolor

Pseudopanax laetus   Wharangi

Tawa [flower]   Rewarewa [juvenile foliage]

White Maire   Rewarewa foliage and flowers

Pokaka *Elaeocarpus hookerianus* **E**
FAMILY ELAEOCARPACEAE
Lowland and montane forest to 1000 m on the three main islands.
Medium-sized canopy tree (H 16 m/TD to 1 m). Juvenile phase, low and sprawling with tier of spreading branches close to ground, second tier near growing point. Leaves small, varied, toothed or lobed. Mature tree, bark grey, rough with small vertical fissures. Leaves narrow, elliptic, serrated. Flowers greenish-white on racemes near branch ends. Fruit capsule-shaped to 20 mm, purplish black.

Horoeka *Pseudopanax crassifolius* **E**
LANCEWOOD
FAMILY ARALIACEAE
Widespread in lowland and lower montane forests on the three main islands, often planted in urban areas.
Medium-sized bushy-headed tree (H 15 m/TD 50 cm) of forest, margins and open ground. Smooth, grey, fluted bark. Juvenile phase has single stem to 3–4 m, leaves very long, up to 1 m, narrow (12 mm), dark green, heavily toothed and downwards pointing, with prominent yellow-orange midrib. As the tree matures a bushy crown develops and the juvenile leaves fall away. Juvenile stage can last up to 20 years. Adult leaves to 200 mm long and 40 mm wide and coarsely toothed. Flowers small, greenish yellow, in many-branched panicles. Fruit globular, dark purple. Commonly planted in urban areas.

Toothed Lancewood *P. ferox* is very similar but smaller (H 6 m), juvenile leaves shorter (500 mm) and more heavily toothed. Mature leaves smaller and less obviously toothed. Similar range but much less common.

Tree Fuchsia *Fuchsia excorticata* **E**
KOTUKUTUKU
FAMILY ONAGRACEAE
Widespread in moist lowland and montane forest on three main islands, especially regenerating forest.
Small forest tree (H 12 m/TD 60 cm), often multi-trunked, twisted and spreading, with distinctive cinnamon-coloured papery bark which is constantly peeling off in long strips. One of very few native deciduous trees. The leaves are alternate, elliptical, pointed and up to 100 mm long, and may be very finely toothed. Flowers are pendant, often directly from trunk or large branch, green and purple becoming red after pollination. May start flowering before leaves appear. Edible berries called 'kononi'. Possibly the largest fuchsia in the world.

Inanga *Dracophyllum longifolium* **E**
FAMILY ERICACEAE
Widespread in coastal, lowland and subalpine scrublands on three main islands.
Small shrub to tall tree (H 12 m) depending on altitude and latitude. Branches slender and erect, leaves long, to 250 mm, needle-like, erect. Flowers small, white, bell-shaped on short sometimes drooping inflorescences.

Ngaio *Myoporum laetum* **E**
FAMILY SCROPHULARIACEAE
Widespread coastal areas and lowland forest in NI and SI, but uncommon in far south.
Small, often bushy tree (H 10 m/TD 30 cm) with rough, furrowed bark. Leaves bright glossy green, long (100 mm), pointed with prominent midrib and oil glands. Flowers white, five petals, with small reddish spots. Fruit purplish-red capsule to 9 mm.

Toru *Toronia toru* **EG**
FAMILY PROTEACEAE
Lowland and montane forest and shrublands south to Bay of Plenty.
Small much-branched tree (H 10 m/TD 20 cm) with smooth grey bark. Leaves alternate, very long, to 200 mm, and narrow, pointed and slightly spathulate. New growth slightly hairy. Flowers four-petalled, yellow to orange on racemes at branch ends. Fruit white or pinkish drupe.

Pokaka [juvenile foliage]   Horoeka, adult

Horoeka, juvenile at right   Tree Fuschia

Inanga flower   Inanga   Ngaio

Toru   Ngaio

TREES AND SHRUBS WITH LONG NARROW LEAVES   **147**

Toro *Myrsine salicina* **E**
FAMILY MYRSINACEAE
Lowland and montane forest, on SI restricted to north and west south to Greymouth.
Small subcanopy tree (H 8 m/TD 60 cm), often with bushy almost lollipop shape, with grey-brown slightly rough bark. Leaves narrow, slightly spathulate and round-ended. Flowers pinkish-white in clusters directly from branchlets, generally below leaves. Fruit small, stalked pinkish berries to 10 mm.

Tainui *Pomaderris apetala* **N**
DOGWOOD, NEW ZEALAND HAZEL
FAMILY RHAMNACEAE
Coastal shrub western NI, also Wellington, Napier, Christchurch and Stewart Is. Frequently cultivated.
Small rather open shrub (H 4 m). Leaves dark green, to 70 mm, heavily veined giving rough appearance, margins rough and slightly recurved, hairy undersides. Flowers have no petals and are pale cream to creamy-yellow in large elongated clusters at branch tip, often profuse.

Kumeraho *Pomaderris kumeraho* **E**
GOLDEN TAINUI, GUM-DIGGERS SOAP
FAMILY RHAMNACEAE
Coastal forest and scrub, northern half of NI.
Similar to Tainui but less heavily veined, hairless leaves, flowers more yellow and in large globular inflorescences. Often seen on roadside banks, or in regenerating bush.

Mountain Mahoe *Melicytus lanceolatus* **E**
MAHOE WAO
FAMILY VIOLACEAE
Lowland and montane forest, and forest margins to 900 m on the three main islands.
Small tree (H 6 m/TD 30 cm) with rather white bark. Leaves alternate, and are long, to 160 mm, narrow, pointed and finely toothed. Flowers small, five petals, purple to yellow-purple, on short stalks directly from branchlets. Fruit dark purple, 4–6 mm in diameter.

Akeake *Dodonaea viscosa* **N**
FAMILY SAPINDACEAE
Shrublands, coastal and lowland forest south to Banks Peninsula.
Small tree (H 6 m/TD 30 cm) with distinctive reddish-brown bark, constantly peeling off in long, narrow strips. Leaves long, to 100 mm, narrow, elliptic, slightly pointed. Flowers have no petals, in dense clusters at ends of branches. Fruit flattened with broad yellowish wings.
Var. *purpurea* has purplish leaves and red seeds.

Poroporo *Solanum aviculare* **N**
FAMILY SOLANACEAE
Widespread coastal and lower montane forest on three main islands, and in disturbed areas, gullies. Common urban weed.
Distinctive open branching shrub (H 4 m). Leaves long, to 250 mm, narrow and pointed, dark, often purplish-green, sometimes semi-pinnate. Flowers large, to 35 mm, rich purple with yellow centres in small hanging bunches. Fruit orange red capsule to 25 mm long.

See also: **Hebes** — Willow-leaved Hebe *Hebe salicifolia*, Koromiko *H. stricta*, Tree Hebe *H. parviflora*; and **Tree Daisies** — Lancewood Tree Daisy *Olearia lacunosa*, Streamside Tree Daisy *O. cheesemanii*.

Tainui

Toro Kumeraho

Akeake Mountain Mahoe

Akeake Akeake flowers Poroporo

TREES AND SHRUBS WITH LONG NARROW LEAVES **149**

TREES AND SHRUBS WITH TOOTHED LEAVES

Takes precedence over pinnate or leathery leaves.

Pukatea *Laurelia novae-zelandiae* E
FAMILY MONIMIACEAE
Lowland and lower montane forest in wet and damp areas, along streams etc. NI, restricted to north and west of SI.
Large emergent forest tree (H 35 m/TD 2 m) with pale grey-brown slightly rough bark. Tall clean trunk except where solitary in farmland when lower branches may regrow. Mature trees have buttress roots and pneumataphores. Branchlets four-sided, leaves rich glossy green, leathery and toothed. Flowers small, greenish-yellow and either unisexual or bisexual. Fruit small, green and pear-shaped producing fluffy windborne seeds.

Kamahi *Weinmannia racemosa* E
FAMILY CUNONIACEAE
Widespread in lowland and montane forest from Waikato and Bay of Plenty southwards, including Stewart Is.
Large canopy tree (H 25 m/TD 1.2 m) with smooth greyish bark. Leaves are dark green, leathery and coarsely toothed. Flowers small, pinkish white, feathery, abundant on racemes up to 110 mm long turning whole tree pinkish-white in summer Fruit small and tufted within seed capsules.

Towai *W. silvicola* is similar but has a juvenile stage with pinnate leaves and white rather than pink flowers. It is found mainly north of Auckland.

Hinau *Elaeocarpus dentatus* E
FAMILY ELAEOCARPACEAE
Widespread in lowland forests of NI and SI.
Medium-sized canopy tree (H 20 m/TD 1 m), bark rough and vertically fissured. Leaves long, to 120 mm, and narrow, slightly toothed, spathulate and distinctly pointed. Longer in juvenile trees. Flowers white, five petals with serrated tips, on racemes at branch ends. Fruit a purple black drupe to 20 mm.

Pigeonwood *Hedycarya arborea* E
POROKAIWHIRI
FAMILY MONIMIACEAE
Widespread in lowland and lower montane forests except in far south.
Medium-sized tree (H 16 m/TD 80 cm) with smooth dark bark. Leaves up to 120 mm long, elliptic, pointed and coarsely toothed. Flowers small, greenish, petal-less, on branching racemes, strongly scented, male and female on separate trees. Fruit red-orange drupe, favoured food of Kereru, New Zealand Pigeon, hence the name.

Ribbonwood *Plagianthus regius* EG
MANATU
FAMILY MALVACEAE
Widespread but irregular on the three main islands, in lowland forests, riverbeds and alluvial terraces.
Medium-sized tree (H 15 m/TD 1 m) with rough, grey bark. Largest deciduous native tree. Juvenile stage is bushy with interlacing branches. Leaves ovate-lanceolate and serrated on petioles up to 30 mm long. Flowers small, greenish yellow, profuse in large drooping panicles. Male yellower, female greener and smaller. Fruit small whitish capsule, one per flower.

Large-leaved Milk Tree *Streblus banksii* E
FAMILY MORACEAE
Coastal and lowland forest south to Marlborough Sounds.
Small subcanopy tree (H 12 m/TD 60 cm) with smooth light brown bark with corky spots. Leaves alternate with prominent veins. Juvenile leaves often deeply lobed, adult slightly toothed. Flowers small greenish-yellow on catkins. Fruit round and red to 6 mm.

Pukatea flowers [buttress roots]　Kamahi

Towai　Hinau

Ribbonwood

Pigeonwood [fruit]　Large-leaved Milk Tree

TREES AND SHRUBS WITH TOOTHED LEAVES　**151**

Tawheowheo *Quintinia serrata* E
FAMILY PARACRYPHIACEAE
Lowland and montane forest south to Taranaki and in the northwest of SI.
Small, slightly bushy tree (H 12 m/TD 50 cm) with smooth, slightly fissured, grey-brown bark. Leaves are alternate, shiny, largely elliptic with wavy margins and slightly toothed. Flowers yellowish-white with five petals bent backwards, on racemes. Seeds held in small brown capsules.

Westland Tawheowheo *Q. acutifolia* has smaller leaves and flowers and is found in subalpine areas in NI and Westland in SI.

Makomako *Aristotelia serrata* E
WINEBERRY
FAMILY ELAEOCARPACEAE
Widespread on the three main islands in lowland and montane forests and forest margins.
Small shrubby tree (H 10 m/TD 40 cm) with dark, slightly furrowed bark. Leaves are opposite, broad, to 90 mm, pointed and heavily toothed, undersides may be slightly reddish. Flowers white to pale red, four deeply lobed petals, in large panicles; male and female on separate trees. Fruit round, to 8 mm, black berries, rather similar to black currants and can be used for wine and jams. An early coloniser of open ground.

Mahoe *Melicytus ramiflorus* E
WHITEYWOOD, HINAHINA
FAMILY VIOLACEAE
Widespread, lowland and montane forest and forest margins on three main islands.
Small often bushy or shrubby tree (H 10 m /TD 60 cm) with smooth pale grey-brown bark. The leaves are alternate, light green, up to 150 mm long and finely toothed. Flowers yellow-green, fragrant, stalked and growing directly from branch. Fruit dark blue to purple 4–5 mm berry.

Kaikomako *Pennantia corymbosa* E
FAMILY PENNANTIACEAE
Lowland and montane forests to 600 m on the three main islands.
Small, slender canopy tree (H 10 m /TD 50 cm) with rough grey-brown bark. Juvenile phase is densely interlaced mass of twigs and branches close to ground. Leaves leathery, shield-shaped and coarsely toothed. Flowers unisexual or bisexual, fragrant, creamy-white in profuse panicles. Fruit black ovoid drupe to 10 mm long.

Lacebark *Hoheria populnea* EG
RIBBONWOOD, HOUHERE, HOUI, WHAUWHI
FAMILY MALVACEAE
Coastal and lowland forest south to Central Plateau. Many cultivars are used in gardens.
Small tree or shrub (H 10 m /TD 50 cm) with smooth green-brown bark becoming mottled with age. Leaves alternate, bright green, ovoid and toothed. Flowers white, scented, profuse, single or in clusters from leaf axils. Fruit winged seeds, five per flower.

Mountain Ribbonwood *H. lyallii* similar to *H. populnea* but found at higher altitudes. The juvenile trees have smaller leaves than the adults and the flowers are larger and whiter than others of the family. One of the few deciduous natives.

Makamaka *Ackama rosifolia* E
FAMILY CUNONIACEAE
Lowland forest in Northland.
Small forest tree (H 12 m/TD 60 cm) with grey and somewhat corky bark. Small twigs and young leaves hairy. Leaves pinnate, light green, leaflets sharply toothed, heavily veined, decreasing in size towards base, up to 10 pairs in young tree decreasing to three to five in mature tree, sometimes reddish underneath. Flowers small, white on much-branched terminal panicles.

Tawheowheo

Makomako [fruit]

Kaikomako

Mahoe [fruit]

Makamaka

Lacebark [flower]

Mountain Ribbonwood

Hutu *Ascarina lucida* **E**
FAMILY CHLORANTHACEAE
Coastal, lowland and montane forest to 750 m on all three main islands, especially
west coast of SI and Stewart Is.
A small tree (H 8 m/TD 30 cm) with rough grey-brown bark. Leaves are bright yellow-
green, heavily serrated on reddish-brown branchlets. Flowers greenish-yellow, male and
female on same panicle. Fruit small bluish drupes.

Whau *Entelea arborescens* **EG**
FAMILY TILIACEAE
Irregularly in coastal and lowland forest in NI and northern end of SI.
Small bushy tree (H 6 m/TD 25 cm) with grey bark, smooth but scarred. Leaves are large,
up to 250 mm long, soft, bright green, almost circular and finely toothed on long stalks.
Flowers are large, white with yellow centres, and grow in large panicles from the leaf axils
near the branch ends. Fruit are very distinctive spiny capsule; this is the only native tree
with this type of seed capsule.

Taurepo *Rhabdothamnus solandri* **EG**
FAMILY GESNERIACEAE
Coastal and lowland forests, stream banks in NI, more common further north.
Small, dense, many-branched shrub (H 2 m) with red-brown stems and twigs. Leaves are
small (30 mm) near circular, coarsely serrated with very prominent veins. Flowers large,
bright red-orange, tubular, similar to nasturtiums. Fruit is a seed capsule 10 mm long.

Marbleleaf *Carpodetus serratus* **E**
PUTAPUTAWETA
FAMILY ROUSSEACEAE
Coastal, lowland and montane forests and forest margins NI, SI and Stewart Is.
Small tree (H to 10 m/TD to 30 cm) with broad, spreading, tiered branches and rough pale
grey bark. Leaves: juvenile small, marbled, toothed leaves on zigzagging branches; mature
narrowly obovate, marbled dark and yellow-green, with margins starting at different levels
on either side of midrib and slightly toothed. Flowers white, six-petalled, star-shaped, in
panicles at the ends of branchlets. Fruit black ridged berry up to 6 mm.

Tree Nettle *Urtica ferox* **N**
ONGAONGA
FAMILY URTICACEAE
Widespread in scrubland and forest margins to 600 m on the three main islands.
Unmistakeable small herb to bushy, many-branched and interlacing shrub (H 2 m).
All but the woody stems densely covered in long, some over 10 mm, stinging hairs or
spines. Leaves long, to 120 mm, narrow, serrated, pointed with large poisonous hairs on
both sides of midrib. Flowers small and green on racemes from leaf axils. Fruit small. A
fearsome plant to be treated with great care as people have been hospitalised after being
stung (one death has been recorded).

Toropapa *Alseuosmia macrophylla* **EG**
FAMILY ALSEUOSMIACEAE
Lowland and montane forest, south to Taranaki and Ruapehu, and then in northern SI.
Small much-branched understorey shrub (H 2 m). Leaves to 150 mm, pointed, lightly
toothed. Flowers long, tubular and red, hanging individually or in small clusters, strongly
scented. Fruit a bright red berry to 10 mm.

See also: **Large Trees Small Leaves** — Tanekaha *Phyllocladus trichomanoides*, Toatoa
P. toatoa, Mountain Toatoa *P. alpines*; **Pinnate Leaves** — Titoki *Alectryon excelsus*; **Large
Trees Small Leaves** — *Nothofagus* spp.; **Small Leaves** — Turepo *Streblus heterophyllus*,
Bush Snowberry *Gaultheria antipoda*; **Palmate or Finger-like Leaves** — Pate *Schefflera
digitata*, Five-Finger *Pseudopanax arboreus*, Mountain Five Finger *P. colensoi*, Houpara
P. lessonii, *P. discolor*, *Raukaua simplex*; **Long Narrow Leaves** — Mountain Mahoe
Melicytus lanceolatus, Pokaka *Eleaocarpus hookerianus*, Rewarewa *Knightia excelsa*, Tree
Fuchsia *Fuchsia excorticata*, Horoeka *Pseudopanax crassifolius*, Toothed Lancewood
P. ferox; **Tree Daisies** — Holly-leaved Tree Daisy *Olearia ilicifolia*, Common Tree Daisy
O. arborescens, Rangiora *Brachyglottis repanda*, *B. hectori*.

Hutu

Taurepo

Whau [flowers and fruit]

Marbleleaf

Tree Nettle

Tree Nettle

Toropapa

TREES AND SHRUBS WITH PINNATE OR SIMILARLY PAIRED LEAVES

Takes precedence over small and leathery, but after long narrow leaves.

Titoki *Alectryon excelsus* E
TITONGI
FAMILY SAPINDACEAE
Coastal and lowland forests as far south as Banks Peninsula and Fox Glacier.
Medium-sized forest tree (H 15 m/TD to 1 m) with fairly short, stout trunk and bushy head. Bark is smooth and grey. Twigs may be furry, leaves pinnate to 400 mm with alternate leaflets, toothed in young leaves becoming smooth margined when mature. Flowers small and deep red or brown on much-branched panicles. Fruit red with black seed held within brown furry capsule.

Kowhai *Sophora microphylla* N
FAMILY PAPILONACEAE
Widespread in lowland forest margins, riverbanks, lakesides on all three main islands, also in gardens and along streets.
Small, slightly drooping, semi-deciduous tree (H 10 m/TD 60 cm) with rough grey or grey-brown bark. Leaves are pinnate up to 150 mm long and with up to 40 pairs of small leaflets. Flowers in small clusters are bright sunshine yellow, tubular and drooping. Flowers early spring. Favoured food of Tui and Bellbirds. Fruit hard seed held in pods up to 150 mm long. Sometimes has a juvenile rather twiggy stage. Seeds very resistant to salt water. New Zealand's national flower.

S. tetraptera is very similar in most respects, but has fewer and larger leaflets, and flower 'keel' protrudes well beyond the wings. Restricted to the east coast of NI from East Cape to Hastings, but commoner in gardens than *S. microphylla*.

Prostrate Kowhai *S. prostrata* is a small shrub (H 2 m) with much shorter pinnate leaves. Flowers smaller with keel extending well beyond the wings, and fewer per cluster.

Tutu *Coriaria arborea* E
TREE TUTU
FAMILY CORIACEAE
Open areas, shrublands, lowland and montane forest margins NI, SI and Stewart Is.
Small shrubby tree (H 8 m/TD 30 cm) with brown somewhat fissured bark. Young trees have smooth bark with small nodules. Stems four-sided. Leaves opposite, short-stalked, ovoid with narrow point. Flowers small, pinkish-yellow on long, to 150 mm, hanging racemes. Fruit, small dark berries. Sap and seeds of Tutu are highly toxic and the source of the glycoside Tutin. A first coloniser.

Small-leaved Tutu *C. pteridoides* is found in Taranaki, but is just one of several similar species with the same common name found in different parts of the country.

Hangehange *Geniostoma lingustrifolium* E
NEW ZEALAND PRIVET
FAMILY LOGANACEAE
Coastal and lowland forest south to north of SI.
Small undershrub (H 3 m) with green somewhat glossy, opposite leaves, while not actually pinnate they can appear so. Flowers are small, greenish-white, lack petals, in clusters on individual stalks directly from branch. Fruit greenish-brown capsule that splits in two.

Kakabeak *Clianthus puniceus* EG
FAMILY FABACEAE
East coast of NI, but uncommon. However, cultivated version is often found in gardens.
A small spreading shrub (H 2 m), leaves are pinnate to 150 mm long, leaflets alternate, up to 15 pairs. Flowers large, bright red, in clusters, petals look like a parrot's beak, hence the name. Fruit in pods up to 80 mm long

See also: **Toothed Leaves** — juvenile only for Towai *Weinmannia silvicola*, Makamaka *Ackama rosifolia*; **Small Leaves** — Scented or Leafy Broom *Carmichaelia odorata*; **Vines** — Supplejack *Ripogonum scandens*, Climbing Rata *Metrosideros* spp..

Titoki [fruit]

Kowhai [flower]

Tutu flowers [fruit]

Kakabeak

Hangehange [flowers]

Small-leaved Tutu

TREES AND SHRUBS WITH PINNATE OR SIMILARLY PAIRED LEAVES   **157**

TREES AND SHRUBS WITH GLOSSY, FLESHY OR LEATHERY LEAVES

Northern Rata *Metrosideros robusta* E
FAMILY MYRTACEAE

Coastal, lowland and montane forest throughout NI, northwest tip of SI only.
A very large tall tree (H 30 m/TD 3.5 m) that normally, but not always, starts life as an epiphyte in the crown of a mature tree, commonly Rimu or other podocarp. Long roots descend to the ground and grow to encircle the host tree, which then dies leaving the Rata freestanding, often with a 'cave' at the base of the trunk. The rata does not appear to strangle or kill the host, but probably hastens its demise. Bark brown, quite thin, falls off in squarish flakes. The Rata has a large bushy crown, leaves to 50 mm are oval with blunt tips, leathery. Flowers dull to bright red, similar to *M. excelsa*.

Southern Rata *Metrosideros umbellata* E
FAMILY MYRTACEAE

Lowland, montane and subalpine forests, uncommon in NI, but widespread SI and Stewart Is.
A medium-sized forest tree (H 20 m/TD to 1 m) with a broad crown, sometimes epiphytic but more often terrestrial. Bark similar to *M. robusta* but may be greyer. Leaves long, to 75 mm, and distinctly pointed. Flowers red with long red stamens turning tree entirely red when in flower.

Pohutukawa *Metrosideros excelsa* E
NEW ZEALAND CHRISTMAS TREE
FAMILY MYRTACEAE

Mainly coastal NI, occasionally inland lakes, south to Taranaki and Gisborne. Introduced to SI and used as an ornamental in urban areas south to Christchurch.
Large, broadly spreading tree (H 20 m/TD 2 m) with short, many-branched trunk with grey, deeply fissured bark that peels off in strips. Older trees may have aerial roots on lower branches. Leaves opposite, leathery, pointed, elliptical, greyish-green, paler and tomentose on undersides, up to 100 mm long. Flowers red and in bunches with profusion of long red stamens turning whole tree red, flowers December–January. Flowers are an important source of nectar for native birds and geckos.

Taraire *Beilschmeidia tarairi* E
FAMILY LAURACEAE

Lowland, lower montane forest south to Coromandel and Waikato, common in Kauri forests.
Large canopy tree (H 20 m/TD 1 m) with smooth, dark brown, often lichen-covered bark. Branchlets covered in red-brown hairs. Leaves elliptic, to 75 mm, tough, glossy dark green above, underside paler, veins prominent with reddish-brown hairs. Flower small, yellow-green on panicles. Fruit large, to 35 mm, oval, reddish-purple – a favoured food of Kereru.

Karaka *Corynocarpus laevigatus* E
KOPI
FAMILY CORYNOCARPACEAE

Coastal and lowland forest south to Banks Peninsula on east coast of SI and Okarito on west coast of SI. Associated with Maori settlements, likely spread south by them as food source.
Medium-sized spreading canopy tree (H 15 m/TD 1 m) with grey bark. Leaves dark green, to 150 mm long, smooth, leathery and elongated elliptic, with recurved margins. Flowers greenish-white, on large panicles, up to 200 mm long, at branch ends. Fruit large, fleshy orange drupe up to 400 mm long. Flesh edible but seed is poisonous unless cooked.

Mangrove *Avicennia marina australasica* N
MANAWA. FAMILY ACANTHACEAE

Tidal creeks and estuaries south to Kawhia and Bay of Plenty.
Medium-sized tree to small shrub (H 15 m/TD 30 cm) with grey, slightly furrowed bark. Breathing roots or pneumatophores evident in surrounding mud. Leaves elliptic, pointed, greyish-green, leathery up to 100 mm long. Flowers small, yellowish in stalked cluster. Fruit large yellowish capsule. Larger in north, very small and shrubby in south of range. Appears to be spreading southwards, likely caused by climate change and increased silting of estuaries due to poor farming practices and coastal development.

Northern Rata [flower]

Southern Rata [flower]

Taraire

hutukawa [flower]

Karaka with fruit

Mangrove [fruit]

TREES AND SHRUBS WITH GLOSSY, FLESHY OR LEATHERY LEAVES   **159**

Tawapou *Pouteria costata* **N**
PLANCHONELLA COSTATA
FAMILY SAPOTACEAE
Coastal forest and shrublands south to Manukau Harbour on west and Gisborne on east.
Medium-sized, closely branched tree (H 15 m/TD 1 m) with rough brown to grey bark.
Leaves to 100 mm and 50 mm wide, but variable, leathery, dark green with much paler
underside. Flower very small and greenish white on short stalk in leaf axils. Fruit bright red
capsule-shaped berry up to 25 mm long.

Broadleaf *Griselinia littoralis* **E**
PAPAUMA, KAPUKA
FAMILY GRISELINEACEAE
Widespread on three main islands in shrublands, lowland to subalpine forest. In moist
areas may be epiphytic with roots to ground.
Broad, spreading tree (H 15 m/TD 1.5 m) with short, rather twisted trunk, bark rough.
Twigs bear leaves up to 120 mm long. Leaves are thick, shiny, broad, blunt, long stalked,
dark, sometimes yellowish-green with yellowish midrib and veins. Flowers unisexual on
separate trees; small, greenish-white on short panicles on woody twigs. Fruit elongated
purplish-black berry up to 7 mm long.

Puka or **Dogwood** *G. lucida* is closely related and similar, normally epiphytic, but sometimes
becoming a freestanding tree. Aerial roots furrowed, main trunk brown and flaky bark.
Leaves larger, thicker, glossier and lopsided at base. Similar range. (See Epiphytes.)

Kawakawa *Macropiper excelsum* **E**
PEPPER TREE
FAMILY PIPERACEAE
Widespread as undershrub and on margins in lowland forest in NI and SI, mainly coastal
on SI to Banks Peninsula in east and southern Westland.
Prolific shrub (H 6 m) with distinctive jointed branches, the leaves are alternate, large and
heart-shaped, 100 mm x 120 mm, and frequently full of holes due to the caterpillar of a
native moth which never eats the whole leaf. Flowers are minute and on small candle-like
stalks which develop into minute orange berries.

THE PITTOSPORUMS – PITTOSPORACEAE
There are some 26 species of *Pittosporum* in New Zealand. They are notable for their small
colourful flowers, with folded back petals, sticky black seeds held in capsules that split
open and smooth-margined alternate leaves.

Black Maipou *Pittosporum colensoi* **E**
Lowland to lower montane forests from Bay of Plenty southwards.
Small tree (H 10 m/TD 40 cm) with spreading branches. Leaves elliptic, pointed, leathery,
prominent midrib and veins. Flowers dark red to purple, singly from leaf axils. Seeds black,
in three-valved capsule, to 12 mm long.

Karo *Pittosporum crassifolium* **E**
Coastal forest margins south to Gisborne; elsewhere cultivated.
Shrub or small tree (H 9 m) with ascending branches. Leaves ovate, thick and leathery
to 100 mm with long, buff, hairy undersides. Flowers rich dark red in small clusters, stalk
and petals hairy, strongly scented. Seed black and shiny in large three-valved capsule
to 20 mm long. Can be mistaken for Pohutukawa which has opposite leaves and very
different flowers. A popular garden plant.

Kohuhu *Pittosporum tenuifolium* **E**
Coastal, lowland and lower montane forests on the three main islands.
Shrub or small tree (H 8 m/TD 30 cm), quite slender in forest locations, very bushy in the
open, with grey blistered bark. Branchlets shiny red-brown. Leaves small, to 40 mm, silvery
green, elliptic, margins wavy, undersides white and hairy. Flowers purple to near black,
singly on short stalk. Seeds black in three-valved capsule to 12 mm long.

Tawapou

Broadleaf [fruit]

Black Maipou [fower]

Kawakawa [maturing flowers]

Karo [flower]

Kohuhu with seed capsule

TREES AND SHRUBS WITH GLOSSY, FLESHY OR LEATHERY LEAVES   **161**

Horopito *Pseudowintera axillaris* **EG**
LOWLAND PEPPER TREE, LOWLAND HOROPITO.   FAMILY WINTERACEAE
Widespread in lowland forest in NI and in the northwest of SI.
Small distinctive undershrub (H to 8 m) with very dark, smooth bark. Leaves elliptic, bright glossy green and peppery when tasted. Flowers small and greenish on woody twigs. Fruit, bright red berries to 6 mm.

THE LARGE-LEAVED COPROSMAS – RUBIACEAE

There are over 50 species of *Coprosma* in New Zealand and many are covered in the small-leaved section. However, those on this page have larger, sometimes glossy leaves with distinctive stipules. The flowers are greenish-white, occasionally purplish; female normally small but with long feathery stigmas, male have long dangling stamens. Berries large and often very colourful. The leaves are strong smelling, especially Stinkwood, hence the generic name 'copros' which is Greek for dung.

Taupata *Coprosma repens* **E**
Coastal areas, shrublands and coastal forest south to northern fringes of SI.
Small tree or shrub (H 8 m) often sprawling, hence its scientific name. Stipules show several black 'teeth'. Leaves dark green, shiny, broadly oval, round-ended and with prominent midvein. Flowers greenish-white, occasionally purplish, female are small and inconspicuous with two long feathery stigmas, male larger with long dangling stamens. Fruit orange-red drupes up to 10 mm long.

Kanono *Coprosma grandifolia* **E**
MANONO, LARGE-LEAVED COPROSMA
Widespread moist lowland and montane forest on the three main islands.
Shrub or small tree (H 6 m). Leaves large, to 200 mm, oval, pointed, pale green or yellow-green, sometimes mottled. Stipules whitish. Flowers similar to *C. repens*. Stipules slightly toothed, greyish, translucent. Fruit bright red shining berries up to 10 mm long.

C. macrocarpa is similar but can be taller and more tree-like. It is found in Northland and as far south as Kawhia and Bay of Plenty. It has smaller, more leathery, leaves; distinct dark papery stipules and very large orange red berries to 25 mm.

Stinkwood *Coprosma foetidissima* **E**
HUPIRO
Higher montane forests and margins from Auckland southwards.
Openly branched shrub or small tree (H 4 m) with brown, rather shiny bark. Leaves thin, light green, oval, round-ended and with small wings on stalk. Flowers similar to *C. grandifolia* but single and female stigmas much longer, up to 20 mm. Fruit orange drupes to 10 mm. Emits smell of rotten eggs when leaves and branches broken or crushed.

Wavy-leaved Coprosma *Coprosma tenuifolia* **E**
Lowland and montane forest, Auckland south to Palmerston North.
Small slender tree or shrub (H 5 m) with upward-pointing branches. Stipules dry and papery. Leaves light green, sometime mottled, pointed and ovoid, stalks may be purplish. Flowers in clusters similar to *C. grandifolia*. Fruit bright orange drupes up to 8 mm long.

Karamu *Coprosma robusta* **E**
Widespread in shrublands and lowland forest on NI and SI.
Small spreading tree or shrub (H 6 m). Stipules are black-tipped. Leaves dark green, leathery, oval and round-ended to 130 mm. Flowers as others. Fruit bright red-orange drupes.

Shining Karamu *Coprosma lucida* **E**
Shrublands, lowland and montane forest on NI, SI and Stewart Is., sometimes as an epiphyte, in drier habitat than *C. robusta*.
Small tree or shrub (H 4 m) with finely furrowed pale bark. Stipules inconspicuous with green points. Leaves dark green, shiny, leathery to 130 mm, ovoid but with small point. Flowers small whitish-green. Fruit bright orange-red drupes.

See also: **Tree Daisies** — Akepiro *Olearia furfuracea*.

Taupata [flower]

Horopito

Kanono [fruit]

Stinkwood

Karamu, flowers [stipe]

Wavy-leaved Coprosma

Shining Karamu

TREES AND SHRUBS WITH GLOSSY, FLESHY OR LEATHERY LEAVES **163**

TREES AND SHRUBS WITH VARIED OR COLOURED LEAVES

Tarata *Pittosporum eugenioides* E
LEMONWOOD
FAMILY PITTOSPORACEAE
Widespread in lowland and lower montane forest NI and SI.
Young trees have a pyramidal shape and brownish, quite smooth bark. Mature trees develop a bushy crown and grey, lightly fissured bark (H to 12 m/TD to 60 cm). Leaves yellowish-green, glossy on top with prominent yellowy midrib, narrowly elliptic to 150 mm with distinct pointed tip and wavy margins, emitting strong scent of lemon when crushed. Flowers pale yellow, in terminal clusters, sweet-scented. Seeds black in two-valved capsule.

Mamangi *Coprosma arborea* E
TREE COPROSMA
FAMILY RUBIACEAE
Lowland forest south to Hawke's Bay.
Small tree (H to 10 m/TD to 40 cm) with brown, rather furrowed bark. Leaves are small, to 60 mm, rounded, pale green and heavily veined giving a mottled appearance, stalk winged. Flowers small, pale greenish-white. Fruit white drupes up to 70 mm long.

Ramarama *Lophomyrtus bullata* EG
FAMILY MYRTACEAE
Coastal and lowland forest south to Nelson and Marlborough.
Small tree or shrub (H to 6 m) with smooth brown bark. Leaves alternate, broadly ovate, mid green and corrugated. In exposed sites leaves may be reddish and very varied. Flowers single in leaf axils, white with profusion of long stamens. Fruit dark reddish-purple to black berry up to 10 mm long.

Mapau *Myrsine australis* E
MAPOU, MATIPOU, RED MATIPOU
FAMILY MYRSINACEAE
Scrublands, lowland and montane forest margins to 900 m on NI, SI and Stewart Is.
Small upright tree or shrub (H to 6 m/TD to 60 cm), bark smooth with corky nodules. Branchlets red-brown. Leaves yellow-green, reddish stalk, elliptic to oblong with small sharp point in younger leaves, wavy margins, to 60 mm by 25 mm. Flowers unisexual, greenish-cream, in small clusters on short stalks directly on branchlets. Fruit small, black, spherical berry or drupe to 3 mm.

Mountain Horopito *Pseudowintera colorata* EG
RED HOROPITO, ALPINE PEPPER TREE, PEPPERWOOD
FAMILY WINTERACEAE
Scrublands, lowland and montane forest and margins to 1200 m on NI, SI and Stewart Is. Not found in Far North.
Small shrub (H to 2.5 m), occasionally tree to 10 m, with slightly furrowed, greenish-grey bark. Leaves alternate on short stalks, elliptic, bluntly pointed, yellow-green, often heavily blotched with red. Flowers white, star-shaped, single or in small clusters, on short stems directly from branchlets. Fruit spherical black berry, up to 5 mm. Peppery taste to leaves and stems.

Alseuosmia pusilla EG
FAMILY ALSEUOSMIACEAE
Mainly montane forest from Mt. Pirongia to Wellington. SI's west coast south to Hokitika.
Small, erect, mainly unbranched understorey shrub (H to 50 cm). Leaves to 80 mm long, light to mid-green with dark red flecks, especially on paler younger leaves. Flowers long, to 10 mm, red, tubular in small clusters directly from stem. Fruit large, to 12 mm, waxy red berry.

See also: **Toothed Leaves** for Marbleleaf *Carpodetus serratus*.

Tarata Mamangi

Mapau Ramarama

Mountain Horopito [leaf] Alseuosmia pusilla

TREES AND SHRUBS WITH VARIED OR COLOURED LEAVES **165**

TREES AND SHRUBS WITH SMALL LEAVES

Kanuka *Kunzea ericoides* E
WHITE TEA TREE, MANUKA-RAURIKI, MANUOEA, TITIRA, ATITIRA. FAMILY MYRTACEAE
Shrublands and montane forests NI and SI, especially well drained soil; pioneer species, especially after fire.
Shrub or bushy tree (H to 15 m/TD to 60 cm) with grey, constantly peeling bark. Leaves small, alternate or in small clusters, narrow, up to 12 mm x 2 mm, pointed, soft, fragrant when crushed. Flowers small, to 5 mm, fragrant, white with purple centre, long stamens, in clusters, often profuse. Fruit dark red-brown capsule up to 4 mm. Divided into three varieties: *K.e. ericoides* in SI, *K.e. linearis* in Northland and *K.e. microflora* in central NI.

Manuka *Leptospermum scoparium* E
RED TEA TREE, KAHIKATOA. FAMILY MYRTACEAE
Widespread on open ground and forest margins to 1000 m on NI, SI and Stewart Is.
Shrub or small tree (H to 4 m) with peeling bark, often black from mould. Leaves small, up to 12 mm by 4 mm, pointed, fragrant when crushed, prickly. Flowers fragrant and profuse, white up to 10 mm, with purple centre and short stamens, individual on short stalks directly off branchlets. Fruit brown capsule up to 10 mm in diameter. Subspecies *L.s. incanum* has pink flowers and slightly hairy leaves, found in far north only. Distinguished from Kanuka by larger flowers and prickly foliage, also never grows as large as Kanuka.

MANUKA	Broader leaves	Big flowers	Prickly to touch
KANUKA	Very narrow leaves	Small flowers	Soft to touch

Turepo *Streblus heterophyllus* E
SMALL-LEAVED MILK TREE FAMILY MORACEAE
Widespread in lowland forest NI and SI.
Small tree (H to 12 m/TD to 60 cm) with rough grey bark and a persistent divaricating juvenile form. Juvenile leaves small, almost round or deeply lobed, with a small point, and finely toothed. Adult leaves alternate, slightly larger, to 25 mm, toothed and occasionally lobed. Flowers very small, greenish white, asexual, on drooping racemes. Fruit spherical bright red berry to 5 mm. Exudes milky sap when twig broken.

Soft Mingimingi *Leucopogon fasciculatus* E
FAMILY ERICACEAE
Shrublands, coastal and lowland forests; often associated with beech forest, also rocky areas south to Canterbury.
Open-branched shrub (H to 5 m). Leaves alternate, lanceolate, pointed, small and very narrow, to 25 mm by 4 mm. Juvenile form may have leaves to 600 mm by 10 mm. Flowers very small, greenish white, fragrant in small drooping panicles. Fruit spherical, deep red or white, up to about 5 mm.

Prickly Mingimingi *Leptecophylla juniperina* subsp. *juniperina* E
FAMILY ERICACEAE
Shrublands, lowland and montane forest NI, SI and Stewart Is.
Shrub (H to 4 m) with dark brown to black bark. Leaves rich green, very fine almost needle-like, up to 15 mm by 1 mm, sometimes larger in moist habitats, veins visible on underside. Flowers small, greenish-white, singly on short stalk near twig ends. Fruit spherical, red or white, to 5 mm.

Weeping Matipo *Myrsine divaricata* E
FAMILY MYRSINACEAE
Widespread shrubland, lowland and montane forest to 1200 m on NI, SI and Stewart Is.
Shrub or small tree (H to 4 m) with weeping, densely divaricating branches and grey bark. Leaves broadly elliptical to 15 mm long, sometimes heart-shaped, quite sparse. Flowers asexual, minute, greenish in small clusters from branches. Fruit spherical, purple or mauve berries, to 5 mm.

Kanuka foliage and flowers

Manuka

Kanuka

Turepo

Soft Mingimingi [flowers]

Prickly Mingimingi [fruit]

Weeping Matipo

TREES AND SHRUBS WITH SMALL LEAVES **167**

Korokio *Corokia buddleioides* **E**
FAMILY ARGOPHYLLACEAE
Coastal and lowland forest and forest margins, south to Rotorua, also widely cultivated.
Small, dense, much-branched shrub (H to 3 m). Leaves narrow and pointed, underside covered in dense white hairs. Flowers bright yellow, star-shaped at tips of branches. Fruit dark red or black berry to 7 mm.

Bush Snowberry *Gaultheria antipoda* **E**
FAMILY ERICACEAE
Widespread in shrublands and forest margins, coastal to alpine, on NI, SI and Stewart Is.
Small erect or spreading shrub (H to 2 m) with bristly branchlets. Leaves alternate, small, clearly veined, toothed, broadly oval to elliptical to 10 mm. Flowers small, white, sometimes red, bell-shaped, individual, in leaf axils. Fruit red or white capsule up to 6 mm in diameter.
Prostrate Snowberry *G. macrostigma* is a sprawling, spreading shrub with narrower leaves and large red-fleshed berries up to 7 mm. Found in alpine and subalpine areas throughout the country.

Teucridium parvifolium **EG**
FAMILY VERBENACEAE
Streamsides and river terraces, lowland forest margins mainly to the east on NI and SI.
Untidy, dense, often straggling shrub (H to 2 m), branchlets square in cross-section. Leaves small, opposite, spear-shaped, flattened petioles as long as leaves. Flowers small, white, individual in leaf axils.

Pinatoro *Pimelea prostrata* **E**
NATIVE DAPHNE
FAMILY THYMELAEACEAE
Widespread NI and SI from montane to subalpine riverbeds, grasslands and scrublands.
Low spreading shrub (H to 50 cm) with small, spear-shaped, pale grey-green, opposite leaves. Creamy-white, strongly-scented, four-petalled flowers form on branch ends, anthers orange. Fruit white and somewhat pointed.
P. pseudolyallii has more pointed, distinctly hairy leaves, especially on new shoots. Found east of the divide in Marlborough, Canterbury and Otago.

Pohuehue *Muehlenbeckia complexa* **N**
FAMILY POLYGONACEAE
Widespread coastal lowlands to montane grasslands and scrublands NI, SI and Stewart Is.
Low, scrambling, sometimes climbing shrub (H to 60 cm, also climbs) that forms dense mass often covering quite large areas, especially on coast. Leaves small, somewhat leathery, rounded with indented tip. Flower small, to 20 mm, greenish-white in clusters. Fruit small, black and glossy but surrounded by succulent remains of flower.

Tauhinu *Pomaderris amoena* **E**
FAMILY RHAMNACEAE
Coastal and lowland scrublands and shrublands south as far as Canterbury.
Many-branched, compact shrub (H to 2 m). Leaves many, short, to 20 mm, narrow, round-ended with recurved margins making them look almost needle-like. Flowers white with recurved petals, with orange-tipped stamens and yellow stigma.

NATIVE BROOMS – *CARMICHAELIA* SPP. FABACEAE
There are more than 20 species of native broom: from very low species such as **Prostrate Broom** *C. apressa*, found on coastal sand dunes, and **Dwarf Broom** *C. nana,* found up into alpine regions to much larger species such as **Weeping Tree Broom** *C. stevensonii* or **Pink Tree Broom** *C. glabrescens*, which can be up to 10 m tall. Many are very restricted in their range, others, such as **Common Broom** *C. australis*, are widespread.

Scented Broom *Carmichaelia odorata* **E**
LEAFY BROOM
Southern NI and northern and northwestern SI.
Bushy, much-branching shrub (H to 3 m) with green, drooping, flattened and serrated branchlets. Leaves very small, to 10 mm, elliptic, opposite and sometimes pinnate. Flowers pink and white, pea-like on small racemes. Seeds in short individual pods.

Korokio   *Teucridium parvifolium*

Bush Snowberry   Prostrate Snowberry

Pinatoro   Pohuehue

Tauhinu   Scented Broom

TREES AND SHRUBS WITH SMALL LEAVES   **169**

Pink Broom *Carmichaelia carmichaeliae* E
LEAFLESS BROOM
Lowland and montane forest, gorges, steep-sided valleys, alluvial terraces northern SI only.
Slender much-branched shrub (H to 5 m) with long leafless branchlets. Flowers mauve-pink, pea-like on large racemes. Seeds in pods, 10 per pod.

NATIVE TREES AND SHRUBS SMALL LEAVES – COPROSMAS
Most of the small-leaved Coprosmas (Family Rubiaceae) are 'divaricating' shrubs where the twigs turn back inwards creating a dense tangled mass of branches. This may have been a protection against browsing birds such as moas. All have very small, opposite leaves, pointed stipules and white or greenish white flowers.

Mingimingi *Coprosma propinqua* E
Streamsides, wetlands and scrub from Mangonui in Northland south to Stewart Is.
Divaricating tree or shrub (H to 6 m) which can be either upright or prostrate. Leaves are 10–14 mm by 2–3 mm, growing in small clusters on the tips of branchlets. Flowers whitish, male in small axillary clusters, female solitary on tip of short branchlets. Fruit is a blue drupe, though whitish before ripening.

Round-leaved Coprosma *Coprosma rotundifolia* E
Coastal, lowland and montane forest along streams and in damper areas, to 600 m NI, SI and Stewart Is.
Large shrub or small tree (H to 5 m) with brown, furrowed and flaky bark, densely spreading or divaricating branches. Leaves bright green, sometimes blotched purple, spade-shaped, sometimes with a small point, and hairy. Fruit bright orange with small crease between two seeds.

Thin-leaved Coprosma *Coprosma areolata* E
Widespread in shrubland, lowland and montane forest NI, SI and Stewart Is.
Divaricating shrub or small tree (H to 5 m). Stipules and petioles hairy. Leaves small, 10 mm across, heavily veined, rounded but with pointed tip. Fruit purple or black, round to 5 mm.

Coprosma crassifolia E
Coastal and lower montane forest and shrublands NI and SI, mainly to the east.
Divaricating shrub or small tree (H to 4 m) with spreading, interlacing branches. Leaves spoon-shaped, mid-green to yellowish-green, veined, leathery, undersides sometimes white, up to 15 mm long. Fruit white or pale yellow berry to 6 mm diameter.

Coprosma rugosa E
Montane and subalpine scrub and grasslands NI, SI and Stewart Is.
Small divaricating shrub or tree (H to 3 m) with smooth brown branchlets. Leaves small, to 10 mm, and narrow. Flowers small, greenish white. Fruit pale blue drupe to 7 mm.

Coprosma colensoi E
Shrubland, lowland and montane forest from Coromandel southwards to Stewart Is.
Small, spreading, diffuse, divaricating shrub (H to 3 m). Leaves very variable, narrow or wider, sometimes with indented tip. Flower small, on short drooping stalk. Fruit orange-red to dark red capsule up to 7 mm long.

Red-fruited Karamu *Coprosma rhamnoides* E
Widespread shrublands, coastal, lowland and montane forest on NI, SI and Stewart Is.
Small, spreading divaricating shrub (H to 2 m), branchlets often downcurved at ends. Leaves either near round or very narrow on the same plant. Fruit small, 3–4 mm, dark red or black berry, the latter commoner in scrubby habitat.

Broad-leaved Coprosma *Coprosma spathulata* E
Lowland forest, south to Gisborne.
Small divaricating shrub (H to 2 m) with small, round, heavily veined leaves indented at the end. Leaf stalk long and winged. Fruit red to near black berry, 6–8 mm long.

See also: **Hebes** — *Hebe* spp.; **Alpines** — various alpine shrubs.

Pink Broom

Mingimingi

Round-leaved Coprosma

Thin-leaved Coprosma [leaf]

Coprosma rugosa

Coprosma crassifolia

Coprosma colensoi

Red-fruited Karamu

Broad-leaved Coprosma

HEBES

There are over 100 species of *Hebe* native to NZ, more than any other plant genus, and most of them are endemic. They have opposite leaves and leaf buds are enclosed by a pair of immature leaves. They are found mainly in open land, shrubland or forest from coastal to alpine areas. Their rounded form and symmetry and variety of their leaves make them popular as hybrids for gardeners. Many species have a very restricted range and no one species is found throughout the country (only three species are found in both NI and SI. Alpine species are generally smaller and can be found in the 'Native Trees and Shrubs – Alpine and Subalpine' section.

Tree Hebe *Hebe parviflora* E
KOROMIKO, TARANGA
Scrub, streamsides and forest margins, lowland and lower montane central and eastern areas from Whangarei south to Marlborough.
Large, bushy, many-trunked shrub (H to 8 m). Bark grey and somewhat knobbly. Leaves lanceolate to 40 mm, pointed, in bunches at the ends of branchlets. Flowers small, white or pinkish, on short, to 30 mm, dense racemes. Seeds held in small capsule.

Willow-leaved Hebe *Hebe salicifolia* N
KOROMIKO, KOKOROMUKA
Lowland and lower montane shrublands, streamsides and forest margins on SI and Stewart Is., apart from Marlborough Sounds.
Large bushy shrub (H to 5 m) with long, to 150 mm, very narrow, pointed leaves, occasionally slightly serrated. Flowers white, occasionally slightly mauve, on long, to 200 mm, drooping racemes. Fruit seed capsules to 35 mm long.

Very similar are *H. paludosa* found in Westland in lowland wetlands; *H. corriganii* found in central NI at higher altitudes and *H. pubescens*, which has hairy twigs and leaf margins and is found on Coromandel Peninsula and Great Barrier and Little Barrier islands.

Cypress Hebe *Leonohebe cupressoides* EG
Rocky outcrops, river margins and scrublands, east of main divide from Marlborough to Otago.
Dense symmetrically shaped shrub (H to 3 m); looks like a very dumpy cypress. Leaves slightly grey-green, very small, pressed close to branches giving twig-like appearance. Flowers white or pinkish-white, in small clusters at branch tips.

Koromiko *Hebe stricta* E
The most widespread hebe. Coastal to montane forests NI and SI. Five varieties: most widespread is **var. *stricta*** which is found throughout NI apart from the southern tip around Wellington; also **var. *atkinsonii*** which is found from Manawatu Gorge south to Marlborough and the northeast coast of NI. Another variety **var. *egmontiana*** is found only on Taranaki. Slender much-branched shrub (H to 2 m) with grey bark. Leaves variable, dark to yellow-green, lanceolate, to 100 mm, occasionally toothed, somewhat pointed. Flowers small, white to lilac or mauve, scented, on racemes to 100 mm. Fruit in seed capsules up to 5 mm long.

Similar species: *H. linguistrifolia* is shorter, up to 1 m, has rather yellow-green leaves with orange midribs. Flowers pale lavender becoming white, found in scrubland and forest margins from Cape Reinga to Whangerei. *H. acutiflora* is similar but found near Kerikeri and in Puketi Forest, Bay of Islands.

Canterbury Hebe *Hebe canterburiensis* E
Tussock grasslands, shrublands and higher altitude beech forests, Tararua Range south, mainly to the west of SI's Main Divide.
Small, compact, spreading, shrub (H to 1 m). Leaves small, to 17 mm, oval, pointed, margins somewhat hairy, in four rows all in line giving symmetrical appearance. Flowers white, in clusters at the tips of the growing spike. Fruit seeds in small capsules.

Similar species: *H. vernicosa* has smooth leaf margins, smaller flowers in larger clusters and is found in beech forests in Marlborough and eastern Nelson.

Tree Hebe

Willow-leaved Hebe

Cypress Hebe

Hebe stricta var. *stricta*

Koromiko

Canterbury Hebe

Hebe vernicosa

Hebe albicans E
Coastal to subalpine open areas in the northwest of SI.
Small, spreading, rather open shrub (H to 1 m). Leaves quite variable, narrowly pointed to ovate, generally slightly fleshy, up to 6 mm long. Flowers white on small terminal inflorescences.

TREE DAISIES
Tree Daisies (Asteraceae) are a common feature of the New Zealand flora and recognizable by their general bushy appearance, the hairy undersides to their leaves, their peeling bark and, when in flower, by their typical composite, daisy-like flowers. The seeds of all species are small, 'hairy' and wind dispersed.

Heketara *Olearia rani* E
Lowland forest south to Nelson and Marlborough.
Small tree or shrub (H to 7 m) with finely furrowed, often peeling bark. Leaves elliptic, slightly pointed, toothed, very prominent midrib and veins on hairy underside. Flowers white with yellow centres, up to 10 mm across, in profusion on large panicles, often completely obscuring foliage.

Rangiora *Brachyglottis repanda* E
Widespread in lowland forests south to Greymouth and Kaikoura.
Small tree or shrub (H to 7 m) with quite smooth grey-brown bark. Leaves large, to 250 mm by 200 mm, on long, to 100 mm, stalks, broadly ovate and with wavy, somewhat indented margins, distinctly paler and hairy on undersides. Flowers small, yellowish, with no discernible petals on large many-branched panicles.

Akeake *Olearia avicenniifolia* E
Open country, shrublands and forest margins to 900 m on SI and Stewart Is.
Small tree or shrub (H to 6 m) with very pale, papery, peeling bark. Leaves elliptic-lanceolate up to 100 mm long, undersides densely hairy and buff-coloured, margins slightly wavy. Flowers white with few petals, on very long stalks, often drooping.

Holly-leaved Tree Daisy *Olearia ilicifolia* E
HAKEKE, HAKEKEKE
Shrublands and forest margins from Pureora in central NI southwards (including Stewart Is), often near treeline.
Small, smooth-trunked tree or shrub (H to 5 m) with long, to 100 mm, rather narrow, leathery, slightly grey-green leaves with wavy margins and prickles. Leaves broader on SI. Flowers white with yellow centres, in large sprays on long stalks.

Similar species: *O. macrodonta* found from sea level to 1200 m from Thames south to Southland. Has rather broader leaves, branchlets and leaf undersides covered in soft white hairs.

Lancewood Tree Daisy *Olearia lacunosa* E
Montane forest and scrubland, Tararua Range south (including Stewart Is).
Small tree or shrub (H to 5 m) with distinctive peeling bark on open branch structure. Leaves very long and narrow, to 170 mm by 25 mm, dark green, midrib and lateral ribs prominently depressed, recurved margins. Flowers small, to 10 mm, and white in large sprays.

Akepiro *Olearia furfuracea* E
TANGURU
Scrublands, forest margins and streamsides to 600 m, south to Ruahine Range–Palmerston North.
Small tree or shrub (H to 5 m) with rough, peeling bark. Branchlets are flattened and grooved. Leaves elliptic, slightly pointed, thick and leathery, 100 mm by 60 mm, dark green and shiny above, buff-coloured and hairy below, margins slightly undulating. Flowers small, white with pale centres on flat topped, branching panicles.

Hebe albicans

Heketara

Rangiora

Akeake

Holly-leaved Tree Daisy [flower] [foliage]

Lancewood Tree Daisy

Akepiro

TREE DAISIES **175**

Common Tree Daisy *Olearia arborescens* E
Shrublands and forest margins from Taupo southwards (including Stewart Is.), often near treeline.
Small tree or shrub (H to 4 m). The leaves are elliptic to ovate, coarsely toothed or indented, pointed, on long stalk, underside grey-green. Flowers white with dark centres, in large sprays on long stalk.

Streamside Tree Daisy *Olearia cheesemanii* E
River banks and forest margins, Coromandel southwards to Nelson Lakes.
Small tree or shrub (H to 4 m) with rough, flaking bark, branchlets grooved and hairy. Leaves lance shaped, pointed, up to 90 mm long, prominent mid-vein on underside. Flowers white and many-petalled with yellow centres, in very large panicles often hiding the foliage.

Coastal Tree Daisy *Olearia solandri* E
Coastal areas from Northland south to Westport.
Small, dense, upright tree or shrub (H to 4 m), with stiff, erect and spreading branchlets. Leaves small, narrow, leathery, short petioles and with revolute margins. Flowers white, individual, on short branchlets.

Brachyglottis hectorii E
Scrublands and forest margins northwest of SI.
Small tree or shrub (H to 4 m). Leaves bright green, to 25 mm long, pointed and with coarsely serrated margins. Flowers white with long narrow petals and yellow centres in large spreading bunches on long stalks.

Muttonbird Scrub *Brachyglottis elaeagnifolia* E
Shrublands and montane forest Bay of Plenty southwards.
Small tree or shrub (H to 3 m) with grooved branches and brown, hairy branchlets. Leaves obovate, to 100 mm, on long hairy stalks, undersides covered in buff-coloured hairs. Flowers yellow, without petals, on panicles up to 150 mm long.

Similar species: *B. rotundifolia* has rather rounder, glossy green leaves with prominent midrib and veins on underside. Larger flower panicles, to 200 mm. Found on SI south of Jackson Bay in Westland and on Stewart Is.

Rock Daisy *Pachystegia insignis* EG
Lowland to subalpine cliffs and rocky places, Marlborough and Nelson.
A handsome, dense, spreading shrub (H to 2 m). Leaves dark green, leathery, white and downy underneath, elliptic, on long petiole, round-ended. Flower large, daisy-like, in profusion on very long stem.

Brachyglottis greyi E
Southeast NI, mainly coastal but inland on cliffs.
Small dense shrub (H to 2 m). Leaves grey-green with prominent white margins and midrib, underside and stems densely hairy. Flowers yellow in compact to diffuse sprays.

Common Tree Daisy

Streamside Tree Daisy

Coastal Tree Daisy

Muttonbird Scrub

Muttonbird Scrub

Brachyglottis hectorii

Brachyglottis greyi

Rock Daisy

TREE DAISIES   **177**

TREES AND SHRUBS – ALPINE AND SUBALPINE SPECIES

Mountain Toatoa *Phyllocladus alpinus* **E**
FAMILY PHYLLOCLADACEAE
Montane and subalpine forest and scrublands from Pureora in the central NI southwards.
Small rather bushy tree (H to 9 m). Leaves on adult trees actually phylloclades (flattened branchlets), grey-green with lobed or broadly toothed margins. Male catkins long, 6–8 mm, female spherical, 6 mm diameter, on phylloclades.

Mountain Pine *Halocarpus bidwillii* **EG**
FAMILY PODOCARPACEAE
Montane and subalpine scrub from Pureora in the central NI southwards.
Small, compact, rounded shrub (H to 3.5 m/TD to 60 cm). Leaves very small, green and pressed onto stem. Male and female cones on same plant, seed black in fleshy, whitish cup.

Mountain Cottonwood *Ozothamnus vauvilliersii* **E**
MOUNTAIN TAUHINU
FAMILY ASTERACEAE
Widespread montane to alpine shrubland and scrub and fellfield on NI, SI and Stewart Is.
Compact, erect shrub (H to 3 m) with grooved branches. Leaves small, dark green, yellow-brown underneath, leathery, slightly pointed, closely packed rather like a hebe. Flowers white in dense, terminal, upwards facing spray.

Thick-leaved Shrub Groundsel *Brachyglottis bidwillii* **E**
FAMILY ASTERACEAE
Alpine and subalpine shrublands and herbfields Raukumara Range southward, excluding Taranaki, NI and SI only.
Small, dense, upright shrub (to 1 m). Leaves thick, dark green above, beige and densely hairy below. Flowers white, apparently petal-less, on short branching sprays.

Snow Totara *Podocarpus nivalis* **E**
FAMILY PODOCARPACEAE
Widespread in alpine and subalpine scrub, shrublands and forest margins from Volcanic Plateau southwards, including Stewart Is.
Low, spreading, dense, much-branching shrub (H from 1–3 m). Leaves dark green, short, leathery, bluntly pointed, thick-margined and spirally arranged. Male inflorescence upright catkins; female on separate plant produces seed on fleshy edible base.

Whipcord Hebe *Hebe* spp. **E**
FAMILY PLANTAGINACEAE
Alpine and subalpine grasslands, shrublands and fellfields NI and SI.
This group of alpine shrubs have the leaves pressed close to the stem so that they appear to be part of it, hence their common name. There are nine species, all quite low (from 100 mm to 1 m high); some are quite localized. All have small white flowers singly or in small clusters at or near the tip of the branches.

Pumice Whipcord *H. tetragona* has four-sided branches and grows to 1 m. It is the only species on NI in most alpine areas. Clubmoss Whipcord *H. lycopodioides* is similar in size but with very slender branches 2–3 mm in diameter and is found in northern and central SI east of the Main Divide. Dwarf Whipcord *H. hectorii* has three subspecies, grows to only 150 mm, while Ochreous Whipcord *H. ochracea* grows to around 30 cm in height, has rather yellowish-green leaves especially at their tips and is found only in the northwest of SI. *H. imbricata* is found only in Central Otago.

The Leonohebes or semi-whipcords are a group of four species, quite similar to the true whipcords, but with less appressed leaves, often giving a square appearance to stems. They are found only in the SI, with very little overlap of ranges. The most widespread of the two are *Leonohebe ciliolata* found in the north and west, and *L. cheesemanii* found mainly to the east of the Main Divide. Some species of the genus *Helichrysum* can look similar to the whipcords.

Mountain Pine

Mountain Toatoa [foliage] [male cone]

Thick-leaved Shrub Groundsel [flower]

Mountain Cottonwood

Snow Totara [flower]

Ochreous Whipcord

Hebe imbricata

Dwarf Whipcord

Dish-leaved Hebe *Hebe treadwellii* E
FAMILY PLANTAGINACEAE

Widespread in SI alpine and subalpine grasslands, screes and rocky areas.
Small, either erect or sprawling, shrub (H to 30 cm). Leaves small, green with yellowing at margins, pointed, somewhat concave. Flowers white in small cluster at branch tip.

Thick-leaved Hebe *Hebe pinguifolia* E
FAMILY PLANTAGINACEAE

Alpine and subalpine rock and fellfield, sometimes grasslands, Nelson south to Otago.
Low, compact shrub (H to 30 cm) with erect branches. Leaves grey-green, thick, cupped. Flowers white in small terminal clusters.

Hebe odora E
FAMILY PLANTAGINACEAE

Widespread from Lake Waikaremoana southward to Stewart Is. in montane and subalpine grasslands and shrublands and in boggy areas.
Low, spreading shrub (H to 30 cm) with erect branches. Leaves bright green, small, slightly cupped. Flowers in small clusters at branch tips.

Hebe buchananii E
FAMILY PLANTAGINACEAE

Alpine and subalpine areas from Aoraki Mt Cook National Park southwards.
Low, spreading, often mat-like with upright branches (H to 30 cm). Leaves broadly elliptical to near circular, fleshy. Flowers white in lateral or terminal inflorescences.

Spreading Grass Tree *Dracophyllum menziesii* E
FAMILY ERICACEAE

Moist alpine and subalpine scrub and grasslands SI and Stewart Is.
Much-branched, low, spreading shrub (H to 1 m). Leaves long, to 200 mm, narrow and pointed, green or red-green, in rosettes at branch ends. Flowers white with red back-curved petals on short compact spray.

Dracophyllum recurvum E
FAMILY ERICACEAE

Alpine and subalpine grasslands and scrublands in central NI.
Spreading rather open shrub (H to 1 m). Leaves back-curving, grey-green to red-green in tufts at branch ends. Smaller narrower leaves and a less dense shrub than *D. menziesii*. Flowers small, white, bell-shaped in small clusters at or near branch ends.

Prostrate Grass Tree *Dracophyllum pronum* E
FAMILY ERICACEAE

Widespread in alpine and subalpine areas NI and SI, mainly to east of divide in SI.
Very low spreading shrub (H to 30 cm) with erect branch ends. Leaves small, needle-like, green to greenish-orange. Flowers white, bell-like with back-curved petals, occur singly on branches.

D. prostratum is similar but smaller and more compact with flowers at branch tips.

Snowberry *Gaultheria depressa* var. *novae-zelandia* E
FAMILY ERICACEAE

Widespread alpine and subalpine grasslands, herbfields, rocky and damp areas on NI, SI and Stewart Is.
Prostrate creeping shrub (to 5 cm). Leaves small, on short petioles, dark glossy green, thick, veined, margins crenulated and often tinged yellow or pink. Flower small, white and barrel-shaped. Fruit a large white, pink or red berry, actually the swollen calyx.

Similar species: *G.d. depressa* has hairy leaves and is not found on Stewart Is.

Dish-leaved Hebe

Thick-leaved Hebe

Hebe buchananii

Hebe odora

Spreading Grass Tree

Prostrate Grass Tree

Snowberry

Dracophyllum recurvum

TREES AND SHRUBS – ALPINE AND SUBALPINE SPECIES   **181**

Mountain Snowberry *Gaultheria colensoi* E
FAMILY ERICACEAE
NI only, alpine and subalpine grasslands from Mt Hikurangi southwards.
Low, sprawling shrub (H to 60 cm) with short, erect branches. Leaves small, rounded and finely toothed. Flowers white, bell-shaped in small terminal spikes. Fruit large red fleshy berry.

Scarlet Snowberry *Gaultheria crassa* E
TALL SNOWBERRY
FAMILY ERICACEAE
Widespread montane to alpine and subnival, scrub, shrublands and rocky areas SI only.
Small, slightly untidy shrub (H to 30 cm). Leaves small, thick, dark glossy green but yellowing with age, elliptic, margins toothed. Flowers white, bell-shaped with red-tipped sepals, in large racemes. Fruit a bright red fleshy berry. Dwarfed at higher altitudes.

Similar species: *G. rupestris* has rather larger leaves and is found mainly in Westland.

Lyall's Speedwell *Parahebe lyallii* E
FAMILY PLANTAGINACEAE
Wet rock and streamsides in subalpine and alpine areas Ruahine Ranges south to Fiordland.
Small, dense, prostrate shrub (H to 30 cm). Leaves small, yellow-green, ovate, slightly fleshy, glossy, clearly toothed. Flowers generally white with pink lines and corolla and yellow throat, in small clusters on long, upright stalks. Fruit seeds held in small capsule.

Shrubby Kohuhu *Pittosporum rigidum* E
FAMILY PITTOSPORACEAE
Subalpine scrub from Central Plateau in NI to Nelson on SI.
Compact, many-branched shrub (H to 30 cm), branches sometimes more evident than leaves. Leaves small, leathery, dark green, elliptic, smooth margined and often 'hidden' within the branches. Flowers dark purple, almost black, with typical *Pittosporum* turned-back petals.

Dwarf Heath *Pentachondra pumila* N
LITTLE MOUNTAIN HEATH
FAMILY EPADICRACEAE
Widespread alpine and subalpine grasslands, shrublands and rocky areas on all three main islands.
Prostrate (H to 80 mm), dense, spreading shrub forming quite large patches. Leaves pointed, elliptic, green often yellow or pink, slightly fleshy. Flowers, star-shaped, white with hairy petals. A large pink fruit often occurring with flowers.

Creeping Pohuehue *Muehlenbeckia axillaris* N
FAMILY POLYGONACEAE
Widespread, montane to alpine open stony ground NI and SI.
Prostrate, spreading shrub (H to 50 mm). Leaves small, green to yellow-green, ovoid, bluntly pointed, slightly fleshy. Flowers small, greenish-white, upwards facing, on short stalk. Seed black, one per flower in translucent/white edible fleshy cup.

Prostrate Coprosma *Coprosma perpusilla perpusilla* E
FAMILY RUBIACEAE
Widespread alpine and subalpine grasslands, shrublands and rocky areas from East Cape southwards to Stewart Is.
A ground-hugging, spreading shrub (H to 30 mm). Leaves very small, green, elliptic to spade-shaped, leathery. Flowers small, greenish-white; female flowers have white styles. Fruit comparatively large red-orange drupe. Often grows intermingled with a variety of other plants, *Pratia*, *Viola,* etc.

Similar species: Two other creeping coprosmas are *C. petriei* which has pale greenish-white drupes and *C. atropurpurea* which has bluish or purple drupes.

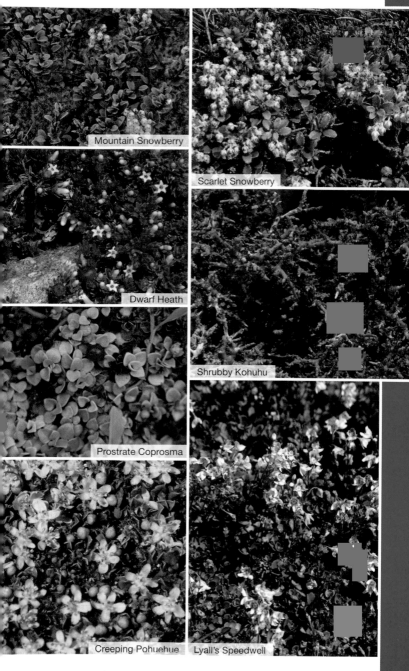

Mountain Snowberry

Scarlet Snowberry

Dwarf Heath

Shrubby Kohuhu

Prostrate Coprosma

Creeping Pohuehue

Lyall's Speedwell

TREES AND SHRUBS – ALPINE AND SUBALPINE SPECIES   **183**

NATIVE VINES, EPIPHYTES AND PARASITES

Vines and epiphytes are well represented in New Zealand, from huge Ratas, which turn into forest trees, to festoons of *Collospermum* and *Astelia* in the crowns of many large trees, to the brilliant red flowers of the native mistletoe. There are also a number of alien vines, several of which are serious pests. (The symbol **E** after the botanical name indicates it is an endemic species and **EG** that it is an endemic genus. **N** = native species. SD = Stem Diameter.)

VINES

Bush Clematis *Clematis paniculata* **E**
PUAWHANGA, PUAWANANGA, PIKIARERO   FAMILY RANUNCULACEAE
Widespread in lowland and lower montane forest, especially margins, NI, SI and Stewart Is.
Attractive climbing vine (SD to 100 mm) reaching up into the canopy. Leaves dark green, glossy, oval, smooth margined or coarsely toothed, in threes on small branchlets; juvenile leaves long and very narrow. Flowers large, to 100 mm, white, six-petalled, in large panicles. Male and female plants, with female flowers smaller. Seeds have long downy hairs.

Yellow Clematis *Clematis foetida* **E**
FAMILY RANUNCULACEAE
Widespread in lowland forests and margins NI, SI and Stewart Is.
Tall scrambling vine (SD to 60 mm) reaching up into the canopy. Leaves dark green, smooth margined or with coarse, rounded teeth; juvenile leaves deeply lobed. Flowers pale yellow, sometimes slightly greenish; male up to 25 mm across, female smaller.

Scented Clematis *Clematis cunninghamii* **E**
FAMILY RANUNCULACEAE
Lowland forest margins from Northland south to northern SI.
Profuse and spreading climber (SD to 40 mm), often covering whole trees. Leaves three-fingered, pointed with long petioles. Flowers off-white to creamy yellow with six long, well separated petals. Seed 'balls' delicate.

Bush Lawyer *Rubus cissoides* **E**
TATARAMOA.   FAMILY ROSACEAE
Widespread in lowland and montane forests and margins on NI, SI and Stewart Is.
Scrambling, very spiny vine (SD to 100 mm) with corky, armband-like swellings on larger stems. Smaller stems and branchlets covered in sharp, reddish, recurved hooks. Leaves variable, lanceolate to elliptic, to 100 mm long and toothed. Flowers off-white to cream in large branching panicles up to 600 mm long. Fruit orange-red, many-seeded berry, similar to introduced blackberry *R. fructicosus*, but smaller.

Similar species: ***R. schmidelioides*** has generally narrower and more coarsely toothed leaves clad with reddish to grey hairs. Margins often rolled downwards, flowers white, fruit yellow. **Swamp lawyer** *R. australis* leaves more rounded, coarsely toothed, flowers white, fruit yellow, prefers damp areas.

Leafless Lawyer *Rubus squarrosus* **E**
YELLOW-PRICKLED LAWYER.   FAMILY ROSACEAE
Widespread on three main islands but less common than *R. cissoides* in lowland and lower montane forest and margins.
Scrambling, very prickly vine (SD to 30 mm). Young stems green with sharp, yellow, recurved hooks. Leaves small, lanceolate, toothed and hooked. Juvenile leaves don't look like leaves, having only midrib. In open, often forms dense, tangled mass of interlocking spiny stems and juvenile leaves on ground. Flowers small and white in panicles. Fruit small orange-red 'blackberries'.

Kiekie *Freycinetia banksii* **E**
FAMILY PANDANACEAE
Widespread in lowland, especially wet forest, on NI and SI, except east and south of SI.
Large grass-like climbing plant (SD to 50 mm) with long, narrow leaves (to 1.5 m by 30 mm)

Bush Clematis seedhead

Bush Clematis   Yellow Clematis

Scented Clematis   Bush Lawyer [spines]

Kiekie   Leafless Lawyer   Bush Lawyer fruit

tapering, pointed and drooping. Flowers small, cream-coloured on finger-like stalks at ends of branches. Male and female on different plants. Fruit like small, hard, green corn cobs. Mature vines drop additional roots down, often forming a dense mass largely obscuring host tree.

Supplejack *Ripogonum scandens* **E**
KAREAO.   FAMILY SMILACACEAE
Widespread in lowland forests on NI, SI and Stewart Is.
Profuse, tangling vine with blackish-brown flexible, jointed stems (SD to 20 mm), often forming dense almost impenetrable confusions. Leaves oval, pinnate-like, on short stalks, clearly veined. Flowers white, small on rather sparse racemes. Fruit bright red berries to 10 mm, leaves often difficult to see.

Pohuehue *Muehlenbeckia australis* **N**
LARGE-LEAVED MUEHLENBECKIA.   FAMILY POLYGONANCEAE
Widespread in lowland and mountain forest especially margins near sea on main islands.
Leafy deciduous vine, often completely covering small trees or shrubs (SD to 5 mm). Leaves thin, oblong to oval, with pointed tip; juvenile leaves violin-shaped. Flowers small, greenish-white in spikes, often quite large. Fruit black, shiny, three-angled enclosed by sepals which are often white. *M. complexa* is similar, more or less deciduous, leaves smaller and thicker, often in small, bushy clumps.

White Rata *Metrosideros perforata* **E**
FAMILY MYRTACEAE
Coastal and lowland forest and margins, south to Banks Peninsula and southern Westland.
Large, heavy-stemmed vine (SD to 150 mm) with dark-brown stringy bark. Leaves small, rounded and almost pinnate especially on young plants, which display a distinctive 'V' shaped pattern as they climb the host's trunk. Flowers white with long white stamens in terminal bunches. Fruit small dark capsule. In open areas, may form a bushy shrub.
   Similar species: **Small White Rata** *M. diffusa* is smaller with slender stems and smaller, more elliptic leaves, slightly pinkish-white flowers and is found throughout the country. *M. colensoi* has rather more pointed leaves which are often overlapping and smaller white or pinkish-white flowers.

Metrosideros albiflora **E**
FAMILY MYRTACEAE
Lowland, especially Kauri forests, south to Bay of Plenty.
Profuse climber (SD to 100 mm), often entirely covering trunk. Leaves dark green, glossy, long, 40–90 mm, and pointed, in pairs, giving pinnate appearance. Flowers white in terminal clusters with long filamentous stamens.
   Similar species: *M. fulgens* has rather smaller, but similarly shaped leaves, bright red flowers and is found throughout NI and in the west of SI. *M. carminea* has smaller, rounder leaves and the red flowers have distinct yellow basal cup.

Kohia *Passiflora tetrandra* **E**
NATIVE PASSION VINE.   FAMILY PASSIFLORACEAE
Lowland forest south to central SI.
Prolific vine (SD to 100 mm) that grows into the forest canopy. Leaves pointed, glossy and with somewhat undulating margins. Flowers greenish-white to 15 mm, with yellow brush-like filaments. Fruit large, to 30 mm, orange.

EPIPHYTES
New Zealand forests abound with epiphytes, with large trees often completely enveloped in them. Notable are the 'nest epiphytes', so called because their large sprays of broad leaves look like untidy nests especially when viewed from below. A number of normally terrestrial trees or shrubs can also be found as epiphytes.

Puka *Griselinia lucida* **E**
SHINING BROADLEAF
FAMILY GRISELINACEAE
Widespread in lowland and lower montane forest on main islands, though commoner on NI.
Small epiphytic tree (SD to 100 mm) with trunk often clasped to, and roots encircling, the

Supplejack [flower]

Pohuehue [flower]

White Rata

White Rata

Metrosideros albiflora

Kohia

Puka

NATIVE VINES, EPIPHYTES AND PARASITES   **187**

host trunk. Leaves large, to 180 mm by 100 mm, rounded, bright green, glossy. Flowers small, greenish in clusters on the twigs. Fruit large, to 10 mm, purple berry.

Kahakaha *Collospermum hastatum* **E**
PERCHING LILY
FAMILY ASTELIACEAE
Coastal and lowland forests south to Nelson and Marlborough.
Large nest epiphyte. Leaves long and narrow, to 1.7 m x 70 mm, with dark bases and arranged in a fan shape. Flowers small and yellow on large, hanging, fingered inflorescences. Fruit small yellow, becoming red, berries.
Similar species: *C. microspermum* is similar but with narrower, more tufted and drooping leaves whose black bases distinguish it from *Astelia solandri*. Found above 300 m on NI only.

Kaiwharawhara *Astelia solandri* **E**
PERCHING ASTELIA, PERCHING LILY
FAMILY ASTELIACEAE
Lowland forest south to Westland, especially moist areas, sometimes on rocks.
Large grass-like epiphyte with long narrow, to 2 m x 40 mm, drooping leaves with silvery grey undersides, forming tufts rather than fans and lacking the dark leaf bases of Kahakaha. Flowers small and red in male or yellow-green in female plants, on large branching racemes, to 500 mm in male, smaller in female plants. Fruit small berries, green becoming yellow-brown.

PARASITES
There are eight species of mistletoe in New Zealand, all but one endemic, the most conspicuous are the beech mistletoes, found mainly on beech trees. They have large colourful flowers and are pollinated by birds and insects. The green mistletoes have small inconspicuous flowers and are insect-pollinated, while the dwarf mistletoes are very small and hard to locate.

Red Mistletoe *Peraxilla tetrapetala* **EG**
FAMILY LORANTHACEAE
Montane and subalpine forest on NI, SI and Stewart Is., especially beech forest.
Conspicuous, parasitic, freely branching quite large shrub, found in the crown of both large and small trees, especially Mountain Beech. Leaves thick, ovate, smooth margined. Flowers large, to 40 mm, bright red in clusters, often profuse giving an amazing splash of colour. Fruit yellow-orange and sticky.
Similar species: **Scarlet Mistletoe** *P. colensoi* is less common, has larger leaves and is found mainly in the south. **Yellow Mistletoe** *Alepis flavida* has smaller yellow flowers and is found only on Mountain Beech. The two green mistletoes *Ileostylus micranthus* and *Tupeia antarctica* are much smaller and have small greenish-yellow flowers and bright yellow fruit, they are found on a variety of host trees.

EPIPHYTIC ORCHIDS
There are eight species of epiphytic orchid, most of them inconspicuous except when in flower.

Easter Orchid *Earina autumnalis* **EG**
FAMILY ORCHIDACEAE
Widespread on trees, occasionally rocks or banks, lowland and montane forests, main islands.
Commonest of the epiphytic orchids. Leaves alternate, long, to 120 mm, and narrow on long, to 1 m, drooping stems. Flowers white with yellow or orange on throat and labellum or lower petal, in large drooping panicles.
Also very widespread is the **Bamboo orchid** or **Peka-a-waka** *Earina mucronata* which has longer, to 200 mm, grass-like leaves and smaller, greenish-yellow flowers, on smaller racemes. **Lady's Slipper Orchid** *Dendrobium cunninghamii* has narrower leaves but with a single large flowers at stem ends. Commonest of the smaller orchids is *Drymoanthus adversus*, with long, pointed, fleshy leaves and conspicuous roots, flowers greenish-white with reddish flecks, in small panicles.

Kaiwharawhara flower | Red Mistletoe

Kaiwharawhara | Kahakaha

Easter Orchid | *Earina mucronata*

NATIVE VINES, EPIPHYTES AND PARASITES   **189**

NATIVE HERBS

New Zealand was originally 80% bush, a habitat that is not ideal for herbaceous plants – plants that have no woody stems and which may be annual, biennial or perennial. It is therefore not surprising to find that most native herbs are found either on the coast, in wetlands, or in the alpine and subalpine areas. While many herb species are habitat specific, others may be found in all areas. (The symbol **E** after the botanical name indicates it is an endemic species and **EG** that it is an endemic genus. **N** = native species.)

NATIVE HERBS – COASTAL AND WETLAND

Lyall's Carrot *Anisotome lyallii* **E**
FAMILY APIACEAE
Coastal cliffs, banks and grasslands to 400 m, Stewart Is. and southern SI.
Large, to 300 mm, tufted perennial herb. Leaves dark green, pinnate with pinnae deeply indented, almost fernlike. Leaf stalks paler. Flowers small, white, in large umbels, smaller on female plants. Seedheads brown.
Similar species: **Haast's Carrot** *A. haastii* is smaller, more compact and with finer leaves, leaf stalks reddish or purple. See Native Herbs – Alpine and Subalpine.

New Zealand Celery *Apium prostratum* **N**
MAORI OR NATIVE CELERY
FAMILY APIACEAE
Coastal rocks and turfs, saltmarshes, seepages and driftwood NI, SI and Stewart Is.
Small, to 100 mm, sprawling, rather variable perennial herb. Leaves glossy green, pinnate, almost bi-pinnate, with deeply lobed pinnae, celery taste. Flowers very small, greenish-white in small umbels.

Yellow Woollyhead *Craspedia uniflora maritima* **E**
FAMILY ASTERACEAE
Rocky coastal areas NI and SI.
Low, to 200 mm, perennial herb. Leaves light green, elliptic, pointed, without petioles, hairy, giving leaves an apparently white margin. Flowers yellow, hemispherical on purplish hairy stalk. Seedhead brown with fluffy windborne seeds.
Similar species: There are at least six species of Woollyhead in New Zealand: *C. uniflora* has four subspecies; **Coastal Woollyhead** var. *pedicellata* has a white flowerhead and is found only on Stewart Is.; **var.** *uniflora* is an alpine species also with a white flowerhead; while *C. incana* is another alpine but with a yellow flowerhead.

Buttonweed *Cotula coronopifolia* **N**
BATCHELOR BUTTONS, YELLOW BUTTONS, WATER BUTTONS
FAMILY ASTERACEAE
Coastal and lowland swamps, sand-hollows, streamsides NI, SI and Stewart Is.
Low, to 200 mm, spreading perennial herb. Leaves light green, fleshy, spathulate, slightly pointed, sometimes partly lobed; leaf base surrounds stem. Flowers yellow, hemispherical, button-like, turning brown.
Similar species: **Soldiers Buttons** *C. australis* has once or twice pinnate leaves, deeply divided, flower stem hairy, flower very pale yellow.

Round-leaved Pincushion *Leptinella rotundata* **E**
FAMILY ASTERACEAE
Quite uncommon, coastal cliffs and boulder falls, west coast of NI.
Small, to 100 mm, spreading, patch-forming herb. Leaves dark green, round, margins serrated, underside hairy. Flower pale yellow, button-like, on short stem.

Shore Groundsel *Senecio lautus lautus* **E**
VARIABLE GROUNDSEL
FAMILY ASTERACEAE
Coastal rocky and grassy areas NI, SI and Stewart Is., found inland as well.
Compact, erect, to 300 mm, bushy annual herb. Leaves dark green, fleshy, heavily and irregularly toothed or lobed with margins rolled under. Flowers bright yellow, seeds fluffy and wind-dispersed.

New Zealand Celery

Lyall's Carrot

Yellow Woollyhead

Buttonweed

Shore Grounsel

Round-leaved Pincushion

Shore Stonecrop *Crassula moschata* **N**
FAMILY CRASSULACEAE
Widespread, coastal, rock, grass, shingle areas, also saltmarsh and seepages, NI, SI
and Stewart Is.
Spreading, decumbent, to 50 mm, mat-forming perennial herb. Leaves green or reddish,
small, fleshy, and pointed on reddish stems. Flowers small, white, four-petalled.

Horokaka *Disphyma australe* **E**
NATIVE ICE PLANT
FAMILY MESEMBRYANTHEMACEAE
Coastal rocky banks and cliffs, shingle beaches, saltmarshes NI, SI and Stewart Is.
Spreading, sprawling, densely branched, succulent herb, height to 100 mm. Leaves green,
sometimes reddish, peg-like, fleshy. Flowers large, all white to bright pink-purple with white
centres and yellow stamens.

Shore Spurge *Euphorbia glauca* **E**
WAIOHAHUKURA, WAIU-O-KAHUKURA, WAIU-ATUA, SEA SPURGE, SAND MILKWEED
FAMILY EUPHORBIACEAE
Coastal cliffs, rocky areas, sand dunes, lakeshore and sheltered coves NI, SI and Stewart Is.
Seriously threatened upright, perennial herb, height 200–400 mm, with multiple individual
stems. Leaves long, bluish-green, pointed, elliptic with no petiole. Stem pale green
becoming red with fruiting, sap milky. Flowers deep reddish-purple, individually from axils
at top of stem. Often part of coastal revegetation projects.

Shore Gentian *Gentianella saxosa* **E**
FAMILY GENTIANACEAE
Coastal areas, rock, turf and sand dunes, Stewart Is. and southern SI.
Small, to 200 mm, clump-forming perennial herb. Leaves green turning yellow with age,
elliptic on winged petioles, pointed and downcurving, on purple stem. Flowers large, white,
upwards pointing in profuse clusters.

Gentianella lineata **E**
FAMILY GENTIANACEAE
Damp, boggy areas, coastal to subalpine SI and Stewart Is.
Very small, low, to 100 mm. Leaves green to brown in small rosette. Flowers bright white,
petals pointed, on individual erect stems, open only in full sunshine.

Small Mudmat *Glossostigma elatinoides* **N**
FAMILY PHYRMACEAE
Wet sand along streams or beside pools NI, SI and Stewart Is.
Very small, to 30 mm, spreading, procumbent herb, often partly covered by sand. Leaves
green, sometimes yellowish or pinkish, fleshy and spathulate, almost oblong. Flowers
small, white with mauve tinge, individual from axils, on very short stem.

Sand Gunnera *Gunnera arenaria* **E**
FAMILY GUNNERACEAE
Coastal areas and estuaries on damp sandy ground, occasionally on sandstone bluffs, NI, SI
and Stewart Is.
Very small, flat, to 40 mm, patch-forming herb. Leaves dark greenish or yellowish-purple,
oval, on short petioles with serrated margins. Flowers very small on short upright spikes.
Fruit small, globular, dull yellow to orange.
     Similar species: **Red-fruited Gunnera** *G. dentata* has green leaves and rather larger
flower spikes with red fruit.

Creeping Gunnera *Gunnera prorepens* **E**
FAMILY GUNNERACEAE
Coastal and lowland damp areas, bogs, lakes and wetlands NI, SI and Stewart Is.
Small, flat, to 50 mm, patch-forming herb. Leaves small, ovoid, dark purplish-green, on
petioles, slightly hairy, margins slightly serrated. Flowers small, white on short upright
spikes. Fruit red, resembling small raspberry.

Shore Stonecrop

Shore Spurge

Horokaka [white flower]

Shore Gentian

Gentianella lineata

Small Mudmat

Sand Gunnera

Creeping Gunnera

NATIVE HERBS – COASTAL AND WETLAND   **193**

Swamp Musk *Mazus radicans* **E**
FAMILY PHYRMACEAE
Coastal, wetland boggy areas, sand dunes, shrubland and scrubland NI, SI and Stewart Is.
Low, to 50 mm, creeping, rooting, perennial herb. Leaves dark green tinged and veined purple, oval with regular dark markings around margin. Flowers violet-shaped, white with yellow lower throat and tongue, upper tongue and petals tinged purple.

Remuremu *Selliera radicans* **N**
CREEPING SELLIERA
FAMILY GOODENIACEAE
Widespread coastal turf, rocky areas, shrublands and saltmarsh, inland streamsides and lakes shores NI, SI and Stewart Is.
Very small, to 20 mm, mat-forming perennial herb. Leaves green, becoming yellow with age, club-shaped, bluntly pointed, fleshy. Flowers white, occasionally pale blue, lopsided with five petals on bottom half of flower only, throat yellow. Fruit small, greenish, fleshy drupe.

Similar species: Could at first sight be confused with **Panakenake** *Pratia angulata.*

Glasswort *Sarcocornia quinqueflora quinqueflora* **N**
SOUTHERN SALT-HORN
FAMILY ARAMANTACEAE
Coastal rocks, banks and saltmarshes NI, SI and Stewart Is.
Low, to 150 mm, spreading, upright, succulent perennial, often forming very large exclusive stands. Leaves dull olive often with pink tinge, leaves fused into shoot which stands erect from spreading stems. Flowers yellow but minute.

Beach Spinach *Tetragonia tetragonioides* **N**
KOKIHI, TUTAE-IKA-MOANA, NEW ZEALAND SPINACH
FAMILY AIZOACEAE
Coastal areas, beaches, dunes, rocky areas, driftwood piles NI, SI and Stewart Is.
Scrambling, sprawling, spreading perennial herb (height to 500 mm). Leaves succulent, rhomboid, pointed. Flowers small, stemless, yellow. Fruit a horned brownish drupe.

Similar species: **New Zealand Spinach** *T. implexicoma* has reddish stems, more deltoid, less succulent leaves. Flowers smaller, on short stems. Fruit a smooth, red, hornless drupe.

Coastal Goosefoot *Chenopodium ambiguum* **N**
FAMILY AMARANTACEAE
Coastal rocks and banks NI, SI and Stewart Is.
Prostrate sprawling herb. Leaves green, coarsely toothed, white undersides, stems often red. Flowers very small, whitish on short spikes.

Rauparaha *Calystegia soldanella* **N**
SHORE BINDWEED, SHORE CONVOLVULUS
FAMILY CONVOLVULACEAE
Coastal or lakeside sandy, pumice, ash or fine gravel areas, occasionally cliffs, NI, SI and Stewart Is.
Spreading, many-branched perennial creeper, often forming dense patches. Leaves thick, green with pale midrib and veins, almost round with petioles. Flower large, pink-and-white striped, trumpet shaped.

Similar species: *C. sepium* is an inland species with similar flower colouration.

Shore Lobelia *Lobelia anceps* **N**
NEW ZEALAND, WILD OR NATIVE LOBELIA
FAMILY LOBELIACEAE
Coastal and lowland, beaches, saltmarshes, rocky areas, riversides, lake shores NI and SI.
Low, spreading, sprawling to erect, perennial herb forming extensive patches. Leaves green, almost strap-shaped, less so at base, pointed with variable petiole, slightly fleshy. Flowers lobed, blue-mauve on opening fading with age.

See also: **Dune oxalis** *Oxalis rubens,* Native Herbs – Lowland and Montane, Yellow Flowers.

Swamp Musk

Glasswort

Remuremu

Beach Spinach

Coastal Goosefoot

Rauparaha   Shore Lobelia

NATIVE HERBS – COASTAL AND WETLAND   **195**

NATIVE HERBS – LOWLAND AND MONTANE

WHITE, GREEN OR PINK FLOWERS

Puatea *Anaphalioides trinervis* **E**
HANGING DAISY
FAMILY ASTERACEAE
Lowland and lower montane banks, streamsides, road-cuttings, NI south to central SI.
Scrambling, rooting, prostrate daisy to 300 mm. Leaves mid-green, short, pointed, elliptic.
Flowers straw-like, white with yellow centre in clusters near end of silvery green, downy
stem, often in profusion.

Papataniwhaniwha *Lagenifera petiolata* **E**
FAMILY ASTERACEAE
Coastal, lowland to subalpine banks and open areas NI, SI and Stewart Is.
Small, to 100 mm, rosette-forming herb, growing in open patches. Leaves green, ovate, on
winged petioles, margins coarsely toothed. Flower white with pale yellow centre.

Similar species: **L. pumila** is larger with dark flower stems and rounder leaves.

Parani *Lagenifera strangulata* **E**
FAMILY ASTERACEAE
Lowland to montane open forest and damp areas, shrublands, dunes NI, SI and Stewart Is.
Small, to 150 mm, rosette-forming herb. Leaves green, round, serrated margins, hairy,
on very long, hairy petioles. Flower very small, white with very short petals, on very tall,
sparely hairy, slender stem.

Bead Plant *Nertera depressa* **E**
FRUITING DUCKWEED
FAMILY RUBIACEAE
Lowland to montane damp open often peaty ground NI, SI and Stewart Is.
Low, to 30 mm, creeping perennial herb. Leaves small, thick, ovoid, pointed, glossy green,
sometimes veined. Flower pale white-green, small and inconspicuous, on very short stem.
Fruit bright red-orange drupe.

Similar species: **N. scapanioides** has slightly larger, circular, hairy leaves on hairy petioles;
Hairy Forest Nertera *N. villosa* has heart-shaped leaves with short, stiff hairs, as does
N. dichondraefolia, which is restricted to northern NI. **N. setulosa** has hairy, pointed,
ovoid leaves while **N. ciliata** is similar but largely hairless and **N. balfouriana** is smaller,
hairless and has pear-shaped fruits.

Lanternberry *Luzuriaga parviflora* **N**
FAMILY LUZURIACEAE
NI only, lowland and lower montane scrub and forest margins, banks and mossy trees.
Delicate, scrambling, to 300 mm, creeping forest herb. Leaves dark green, long, narrow,
pointed, with thickened margins. Flowers white, bell-shaped. Fruit a white, pointed drupe.

Parataniwha *Elatostema rugosum* **E**
FAMILY URTICACEAE
Lowland and montane forest, damp shady places where it is the dominant ground cover NI,
SI and Stewart Is.
Spreading, leafy perennial, to 300 mm, forming dense exclusive patches. Leaves purple
becoming dark bronze-green with purple midrib and veins, pointed ovate-elliptical, toothed
margins. Flowers greenish white, inconspicuous.

Puatea

Papataniwhaniwha

Parani

Lanternberry

Bead Plant

Parataniwha

Hairy Forest Nertera

NATIVE HERBS – LOWLAND AND MONTANE, WHITE, GREEN OR PINK   **197**

White Oxalis *Oxalis magellanica* **N**
FAMILY OXALIDACEAE
Moist lowland and lower montane streamsides and shady banks, NI, SI and Stewart Is.
Low, to 100 mm, spreading herb forming spare or dense patches. Leaves green, trifoliolate on short stems or petioles, slightly hairy underside. Flower individual, white with green throat.

Rengarenga *Arthropodium cirratum* **E**
ROCK LILY, RENGA LILY   FAMILY LAXMANIACEAE
Lowland and coastal drier areas, south to Nelson.
Tall, to 1 m, elegant, tufted perennial herb. Leaves dark green, long, narrow, pointed, drooping. Flowers white to purplish-white in large drooping racemes on tall, dark stems.

Similar species: *A. bifurcatum* is very similar, often larger with broader leaves and larger inflorescences, commonly cultivated.

Onion Orchid *Microtis unifolia* **E**
FAMILY ORCHIDACEAE
Open grasslands and herbfields, lowlands to subalpine NI, SI and Stewart Is.
Tall, to 700 mm, slender orchid. Leaves long and slender, one per flower spike, attached halfway up spike. Flowers small, greenish white, on long, slender stalk, labellum more or less rectangular.

Similar species: **Small Onion Orchid** *M. oligantha* is similar but much smaller with only a few flowers per spike. *M. parviflora* has a triangular labellum which looks like a down-curved tongue. The **Leek Orchid** *Prasophyllum colensoi* has purplish flowers.

Maikuku *Thelymitra longifolia* **N**
MAIKAIKA, WHITE SUN ORCHID   FAMILY ORCHIDACEAE
Wide range from coastal to subalpine, especially shrublands NI, SI and Stewart Is.
Small, to 150 mm, delicate perennial herb, sometimes in dense colonies. Leaf is long and narrow, to 20 mm, down-curved, slightly fleshy leaves halfway up stem. Flower white sometimes pale pink, with delicate parallel vein-like lines.

Similar species: Variable and may be several distinct species. There are 12 other species of this genus in NZ, mostly with blue or mauve flowers. **Colenso's Sun Orchid** *T. colensoi* has blue/mauve flowers, while the **Blue Sun Orchid** *T. pauciflora* has a blue flower.

Greenhood *Pterostylis banksii* **E**
TUTUKIWI   FAMILY ORCHIDACEAE
Widespread on lowland to lower montane forest NI, SI and Stewart Is.
Graceful, distinctive upright orchid, to 600 mm. Leaves light green, grass-like on flower stalk. Flower pale green with faint white lines, hooded with two long ears or horns.

Rauhuia *Linum monogynum* **E**
NEW ZEALAND TRUE OR LINEN FLAX   FAMILY LINACEAE
Sheltered coastal and lowland banks and cliffs NI, SI and Stewart Is.
Upright, to 500 mm, rather bushy perennial herb, base of stems somewhat woody. Leaves bluish-green, narrowly elliptic, pointed, and pointing upwards. Flowers large and white at end of stem.

Similar species: **Pale Flax** *L. bienne*, an introduced species, has pale blue flowers.

Pink Bindweed *Calystegia sepium* subsp. *roseata* **N**
AKAPOHUE, POHUE   FAMILY CONVOLVULACEAE
Coastal and lowland scrub and forest, and wetland margins, urban areas NI, SI and Stewart Is.
Profuse, spreading, climbing, perennial herbaceous vine. Stems dark. Leaves green to yellow-green, deltoid to ovate with petioles. Flower large pink-and-white trumpet.

Similar species: **New Zealand Bindweed** *C. tuguriorum* has rather larger all-white flowers, while the **Small-flowered White Bindweed** *C. marginata* has narrower, more pointed leaves and much smaller flowers. **Great Bindweed** *C. sylvatica* is much larger, has all-white flowers and has large inflated green bracts covering the much smaller sepals.

White Oxalis

Onion Orchid Rengarenga

Maikuku Greenhood

Rauhuia Pink Bindweed

NATIVE HERBS – LOWLAND AND MONTANE, WHITE, GREEN OR PINK   **199**

YELLOW FLOWERS

Hairy Buttercup *Ranunculus reflexus* **E**
MARURU, KOPUKAPUKA, PIRIKAHU
FAMILY RANUNCULACEAE
Lowland, montane and subalpine forest floors, scrub and shrubland NI, SI and Stewart Is.
Small, to 300 mm, tufted perennial forest herb. Leaves green, hairy, trilobed, sometimes completely divided. Flower small, yellow with five well-spaced petals on tall, dark stem, sometimes branched with occasional leaves on stem. Seedhead small spiky sphere.

Similar species: **Grassland Buttercup** *R. multiscapus* has more rounded, coarsely toothed leaves, flower larger on shorter stem. Seedhead has small hooks.

Fern-leaved Pincushion *Leptinella squalida* **E**
FAMILY ASTERACEAE
Coastal to subalpine grasslands and scrub NI and SI, often in gravel or sand by rock outcrops.
Low, to 100 mm, creeping, patch-forming perennial herb. Leaves green, sometimes coppery-brown, pinnate, pinnae may be toothed and overlapping. Flower pale yellow, yellow-green, sometimes almost white, on short stalk

Similar species: *L. calcarea* has rigidly stiff, fleshy leaves lacking any brown pigmentation and is found only in SI. *L. serrulata* has silvery hairy leaves, pinnae more oblong than triangular.

Fireweed *Senecio hispidulus* **N**
FAMILY ASTERACEAE
Lowland to montane open forest and disturbed and urban areas NI, SI and Stewart Is.
Tall, to 1 m, flowering, short-lived annual herb. Leaves dark purplish green, deeply and irregularly serrated in basal rosette. Flowers small, yellow, petal-less in large, disparate umbel. Seedhead fluffy. Seeds windborne.

Similar species: There are several similar and related species, mostly introduced.

Wharanui *Peperomia tetraphylla* **N**
FAMILY PIPERACEAE
Coastal and lowland forest, often on bank or tree, streamsides from Northland to Bay of Plenty and East Cape.
Bushy, to 300 mm, much-branched, spreading, epiphytic or rupestrial, perennial herb. Leaves glossy green, whorled or opposite, slightly fleshy, ovoid, bluntly pointed, midrib and two parallel lateral veins clearly visible. Flowers minute greenish-yellow on long, to 100 mm, terminal inflorescence.

Similar species: *P. urvilleana* is larger, leaves alternate, longer and more pointed, inflorescence axillary as well as terminal, also more widespread, including northern SI.

Creeping Oxalis *Oxalis exilis* **N**
YELLOW OXALIS
FAMILY OXALIDIACEAE
Widespread lowland and montane shrubland and scrubland, disturbed and urban areas NI, SI and Stewart Is.
Spreading, patch-forming rather delicate herb to 200 mm high. Leaves light to bright green, trifoliolate, on long, slender stalks or petioles. Flowers yellow on slender stalk, one or several per stalk.

Similar species: **Dune Oxalis** *O. rubens* is a coastal species with individual plants commonly growing through other vegetation.

Hairy Buttercup

Wharanui

Fern-leaved Pincushion

Fireweed

Creeping Oxalis

NATIVE HERBS – LOWLAND AND MONTANE, YELLOW   **201**

NATIVE HERBS – ALPINE AND SUBALPINE

As befits a mountainous country, New Zealand has a large and varied alpine flora. Nearly half of all native plant species are found in the alpine and subalpine regions which range from around 700 m to 2000 m, depending on latitude. Much of this habitat is in the SI, but it is also found around the NI volcanoes and along the Tararua, Ruahine and Raukumara ranges. Some species described here are also found in montane and lowland areas, but are more likely to be found in alpine and subalpine areas. Heights given are very much dependant on altitude, the more alpine the climate, the smaller the plants tend to be.

WHITE FLOWERS

Giant Speargrass *Aciphylla colensoi* E
TARAMEA, WILD SPANIARD
FAMILY APIACEAE
Alpine and subalpine grasslands from East Cape to Canterbury.
Large tufted plant with dramatic flowerspike to 2 m. Leaves rigid, narrow, pointed, once or occasionally twice pinnate, slightly bluish-green with yellow or orange midrib. Flowers greenish-white with leafy orange-coloured bracts on a tall spike with long, to 150 mm, very sharp spines.

Common Speargrass *Aciphylla squarrosa* E
TARAMEA, SPANIARD
FAMILY APIACEAE
Alpine and subalpine grasslands NI only.
Large tufted plant to 1.5 m. Leaves long, narrow, pointed, slightly serrated, bluish green and up to three times pinnate. Flowers yellow-white on large upright flowerspike with very long, to 300 mm, spines normally downward pointing.

Pygmy Speargrass *Aciphylla monroi* E
LITTLE SPEARGRASS
FAMILY APIACEAE
Widespread in alpine and subalpine grasslands, herbfields and rocky areas Nelson to North Canterbury.
Short, compact, tufted plant to 250 mm. Leaves green to yellow-green, narrow, pointed and divided into 2–6 pairs of leaflets. Flowers off-white in divided umbel. Seedheads brown and spherical.

Kopoti *Anisotome aromatica* E
AROMATIC ANISEED
FAMILY APIACEAE
Widespread in montane to alpine grassland and shrubland and scrub NI, SI and Stewart Is.
Very small to small, 100–500 mm, umbellifer. Leaves small, delicate, pinnate, irregular or toothed, often flattened and hidden in grass. Flowers creamy white, umbel spread out and irregular. Seedheads turn brown.

Bristly Carrot *Anisotome pilifera* E
FAMILY APIACEAE
Rocky areas, subalpine to snowline SI only.
Medium-sized umbellifer to 600 mm. Leaves pinnate, bluish-green and coarsely toothed, or dark green and more finely toothed. Flowers white in large dense umbels. Stems sometimes purplish. Seedheads turn brown.

Haast's Carrot *Anisotome haastii* E
FAMILY APIACEAE
Widespread in montane and subalpine scrub and shrublands, also alpine grasslands SI and Stewart Is.
Medium-sized umbellifer to 600 mm. Leaves dark green, pinnate and finely toothed. Flowers white on large, somewhat spreading umbels. Stalks have fine purple stripes. Seedheads turn brown.

Giant Speargrass

Common Speargrass

Pygmy Speargrass   Kopoti

Bristly Carrot   Haast's Carrot

NATIVE HERBS – ALPINE AND SUBALPINE, WHITE FLOWERS   **203**

New Zealand Angelica *Gingidia montana* N
MAUPIRO, MOUNTAIN ANISE
FAMILY APIACEAE
Montane to alpine rocky areas, shrublands, grasslands NI, SI, Stewart Is.; prefers damp places.
Small, to 500 mm, compact umbellifer. Leaves dark green, sometimes glossy, pinnate and finely toothed. Flowers white in somewhat diffuse umbels. Seedheads turn brown and have strong aniseed taste.

Mount Cook Lily *Ranunculus lyallii* E
FAMILY RANUNCULACEAE
Upper montane to alpine grasslands, herbfields from Lewis Pass south, including Stewart Is.
Very large, to 1 m, attractive upright buttercup. Leaves large, to 400 mm, dark glossy green, almost circular with somewhat serrated margins. Flowers large, white with yellow centres, in small spray on tall upright stem. Seeds light green in tight clusters.

Common Gentian *Gentianella bellidifolia* E
SMALL SNOW GENTIAN
FAMILY GENTIANACEAE
Widespread in alpine and subalpine grasslands NI and SI.
Small, 50–300 mm, perennial tufted plant. Leaves glossy green somewhat spathulate. Flowers white on short lateral stems, one or two per stem.

Mountain Gentian *Gentianella montana* E
FAMILY GENTIANACEAE
Alpine and subalpine grasslands and herbfields NI and SI.
Tall, to 500 mm, distinctive gentian. Leaves dark green in basal rosette and on flower stem. Flowers white, upward facing, in large cymes.
　　Similar species: **Alpine Gentian** *G. patula* slightly smaller and more flower spikes per plant. Flowers, especially at lower altitudes, have purplish colouring and veins on corolla.

Snow Gentian *Gentiana corymbifera* E
TALL GENTIAN
FAMILY GENTIANACEAE
Widespread montane to snowline grasslands, shrublands, open ground NI, SI and Stewart Is.
Tall, to 600 mm, distinctive grassland plant. Leaves in basal rosette and on flower stem, dark green, long and pointed. Flowers white in large clusters or cymes, stems greenish purple. Higher altitude form has broader leaves.

Tutumako *Euphrasia cuneata* E
NEW ZEALAND EYEBRIGHT
FAMILY OROBANCACEAE
Montane to alpine grasslands, herbfields, shrublands and rocky areas NI and SI.
Shrubby perennial herb to 600 mm. Leaves small, dark green, crenulated and slightly fleshy. Flowers abundant, white with yellow throat and fine purple stripes.

Similar species: *E. townsonii* is a low, spreading perennial with pointed leaves restricted to northwest Nelson; *E. revoluta*, a low annual herb, is similar but with pointed leaves with the margins rolled under giving a darker appearance. It has a downy calyx with purplish edges. *E. zelandica* has hairy recurved leaves, and the **Tararua Eyebright** *E. drucei* is a small, tufted, creeping perennial herb that prefers damp and boggy places.

Alpine Cushion *Donatia novae-zelandia* E
FAMILY STYLIDACEAE
Alpine bogs and damp herbfields, Tararua Ranges southwards, including Stewart Is.
Distinctive spreading cushion plant, to 30 mm. Leaves dark green and very short in compact rosettes. Flowers white, unstalked in centre of each leaf rosette.

Rock Cushion *Phyllachne colensoi* N
FAMILY STYLIDACEAE
Widespread alpine and subalpine herbfields, fellfields and rocky places NI, SI and Stewart Is.
Low, to 100 mm, spreading, mat plant. Leaves short, to 5 mm, and bluntly pointed. Flowers small, white, five-petalled with protruding stamens amongst the leaves.

New Zealand Angelica

Mount Cook Lily

Common Gentian

Tutumako

Mountain Gentian

Alpine Cushion

Snow Gentian

Rock cushion

Euphrasia revoluta

Alpine Forget-me-not *Myosotis suavis* **E**
FAMILY BORAGINACEAE
SI, rock ledges and crevices, alpine to snowline.
Small, to 200 mm, tufted perennial herb. Leaves pale green, downy, ovate and spathulate, distinct midrib especially at base. Flowers small, white with pale yellow centre in small, dense, terminal cluster. There are several other species of native Forget-me-not with brown, white or yellow flowers.

North Island Forget-me-not *Myosotis eximia* **E**
FAMILY BORAGINACEAE
Alpine, subalpine, higher montane grasslands, scrublands; Coromandel south to Wellington.
Attractive upright herb to 300 mm. Leaves greyish-green, elliptic with small distinct point, on long winged petioles, slightly downy. Flowers white on tall, unfurling spray.

South Island Mountain Foxglove *Ourisia macrocarpa* **E**
SNOWY MOUNTAIN FOXGLOVE
FAMILY PLANTAGINACEAE
SI montane to alpine grasslands and herbfields, especially damp areas.
Distinctive perennial herb, to 700 mm, with creeping rhizome. Leaves dark green, sometimes purple beneath, pointed ovate and bluntly toothed on hairy petioles. Flowers white with yellow centre in large terminal sprays. Capsule in calyx. Two varieties: var. *calycina* found north of Franz Josef and var. *macrocarpa* to the south.

North Island Mountain Foxglove *Ourisia macrophylla* **E**
FAMILY PLANTAGINACEAE
NI alpine and subalpine herbfields and grasslands.
Very similar to *O. macrocarpa*; slightly smaller, to 600 mm, with more rounded leaves.

Creeping Ourisia *Ourisia caespitosa* **E**
FAMILY PLANTAGINACEAE
Widespread, montane to snowline, damp and rocky areas NI, SI and Stewart Is.
Small prostrate herb to 120 mm. Leaves glossy green, lobed, slightly succulent look, in small rosettes. Flowers white with yellow centre on short stems, one or two per stem.

Alpine Avens *Geum uniflorum* **E**
FAMILY ROSACEAE
SI alpine and subalpine herbfields, rocky and damp areas, Nelson south to Otago.
Prostrate plant to 50 mm forming large patches. Leaves dark green to purple, almost circular, margins irregular. Flowers large, white with dark centre, on individual stems.

Alpine Sundew *Drosera arcturi* **N**
FAMILY DROSERACEAE
Montane to alpine bogs NI, SI and Stewart Is.
Small, tufted, insectivorous plant to 150 mm, often forming large patches. Leaves green or reddish-brown, round-tipped and parallel-sided with long, sticky hairs all over. Flowers white, very small on thin hairless stem, one flower per stalk.

Sundew *Drosera spatulata* **N**
FAMILY DROSERACEAE
Widespread in coastal to alpine wetlands NI, SI and Stewart Is.
Very small, to 80 mm, ground-hugging plant. Leaves red, circular on long, wide petioles and covered with long, sticky hairs. Flowers white in small cluster on thin stem, rarely seen open.

Similar species: *D. stenopetala*, or **Wahu**, is normally larger, has longer petiole, spoon-shaped leaves and only one flower per stem.

Scree Epilobium *Epilobium pycnostachyum* **E**
FAMILY ONAGRACEAE
Alpine and subalpine screes and rocky ground NI and SI.
Low, to 250 mm, spreading, scree-loving plant, with deep taproot. Leaves reddish green, elliptic, somewhat pointed and rough-margined. Flowers white from leaf axils at stem ends.

Alpine Forget-me-not Alpine Sundew

Alpine Avens Creeping Ourisia

South Island Mountain Foxglove North Island Mountain Foxglove

Sundew Scree Epilobium North Island Forget-me-not

NATIVE HERBS – ALPINE AND SUBALPINE, WHITE FLOWERS **207**

Glossy Willowherb *Epilobium glabellum* **E**
FAMILY ONAGRACEAE
Localised, alpine and subalpine rocky areas, slips and screes NI, SI and Stewart Is.
Small, to 250 mm, tufted, erect herb. Leaves reddish green, deeply toothed. Flowers white or pink.

Short-flowered Cranesbill *Geranium brevicaule* **E**
NAMUNAMU, *GERANIUM SESSIFLORA* VAR *GLABRA*
FAMILY GERANIACEAE
Widespread, montane to alpine short grassland, open ground, rocky areas NI, SI, Stewart Is.
Small, to 50 mm, prostrate herb. Leaves green to bronze, lobed, sometimes hairy. Flowers white, single, on short stalk. Capsule beak-shaped.

Panakenake *Pratia angulata* **E**
PINAKITERE
FAMILY LOBELIACEAE
Widespread, montane to subalpine grasslands, scrublands, open forest NI, SI, Stewart Is.
Low, to 50 mm, prostrate, creeping herb, forming dense patches. Leaves green, sometimes purple below, alternate, irregularly shaped. Flower white, five petals all on one side of flower. Fruit a purple drupe.

New Zealand Violet *Viola cunninghamii* **E**
WHITE VIOLET
FAMILY VIOLACEAE
Widespread, montane to alpine grassland, shrubland and open scrub NI, SI and Stewart Is.
Small, to 100 mm, tufted herb. Leaves dark green, spade-shaped, softly toothed. Flowers white with orange centre and purple stripes.

Penwiper *Notothlaspi rosulatum* **E**
FAMILY BRASSICACEAE
SI, uncommon and dispersed, on alpine screes from Marlborough to Canterbury.
Unusual and distinctive fleshy herb, flower to 250 mm. Leaves flattened rosette of dark grey-green, on petioles, toothed, overlapping, hairy. Flower white, on stout conical inflorescence. Flowers once in second or third year, seeds and dies.

Bidibidi *Acaena anserinifolia* **E**
BIDIBID, BIDDY BIDDY, PIRIPIRI
FAMILY ROSACEAE
Widespread lowland to subalpine grasslands NI, SI and Stewart Is.
Scrambling, spreading, dense cover plant to 300 mm. Leaves greyish-green, pinnate, toothed margins. Inflorescence spherical on tall stem, petals not visible, anthers white. Seedhead starts green becoming brown or reddish brown, seeds barbed and love woolly socks!

Similar species: 20 species or subspecies of *Acaena* in NZ: *A. novae-zelandia* and *A. microphylla* are known as **Red Bidibidi** due to their bright red seedheads; **Blue Mountain Bidibidi** *A. inermis* is distinguished by dark blue-grey, almost slate-coloured leaves.

False Buttercup *Schizeilema haastii* **E**
FAMILY APIACEAE
Alpine to subalpine rocky areas NI, SI.
Small, to 200 mm, buttercup-like plant. Leaves dark green, glossy, almost circular, with irregular margins on petioles. Flowers in small umbels partially hidden within the leaves.
Var. *haastii* has greenish-yellow flowers and **var.** *cyanopetalum* has purplish flowers.

Kelleria dieffenbachii **N**
FAMILY THYMELAEACEAE
Alpine and subalpine, herbfields and fellfields NI and SI.
Very small, to 70 mm, spreading perennial, looks like small whipcord hebe. Leaves very small and appressed to stem. Flowers minute, white, in small cluster at stem tips.

Glossy Willowherb

Short-flowered Cranesbill

Panakenake

New Zealand Violet

Kelleria dieffenbachii

Bidibidi

Penwiper

False Buttercup

NATIVE HERBS – ALPINE AND SUBALPINE, WHITE FLOWERS **209**

COLOURED FLOWERS

Maori Onion *Bulbinella hookeri* **E**
GOLDEN STAR LILY, BULBINELLA   FAMILY ASPHODELACEAE
Montane and subalpine grasslands and shrublands NI and northern part of SI.
Prominent tufted plant, to 600 mm, with tuberous rootstock. Leaves green, long, narrow, pointed, slightly fleshy, dying down in winter. Flowers yellow conical raceme on tall spike.

Similar species: *B. angustifolia* has narrower leaves and is found east of SI; *B. gibbsii* var. *balanifera* in central and western Southern Alps; *B. gibbsii* var. *gibbsii* on Stewart Is. and *B. talbotii* only in Gouland Downs, Kahurangi NP. All are very similar in appearance.

Vegetable Sheep *Raoulia eximia* **EG**
COMMON OR RED-FLOWERED VEGETABLE SHEEP   FAMILY ASTERACEAE
Rocky and exposed alpine areas to 1800 m, SI only.
Large, to 600 mm and 1–2 m wide, rounded, cushion plant. Leaves very short, in rosettes, bluish green with white hairs. Flowers red, small, in centre of each leaf rosette.

Similar species: Giant Vegetable Sheep *Haastia pulvinaris* has buff hairs and yellow flowers; Red-flowered Vegetable Sheep *R. rubra* is much smaller, to 250 mm across; Silvery Vegetable Sheep *R. mammillaris* has a very restricted range and yellowish flowers.

Rimuroa *Wahlenbergia violacea* **E**
HAREBELL   FAMILY CAMPANULACEAE
Widespread coastal to subalpine grasslands NI, SI and Stewart Is.
Small, delicately flowered herb. Leaves green or reddish-green, somewhat spathulate, slightly toothed in flattish rosettes. Flower purple to very pale flax blue or even white, on long, slender green or dark stem (to 200 mm), somewhat bell-shaped when just opened. Very similar to *W. albomarginata*, which is found only in alpine and subalpine areas.

New Zealand Harebell *Wahlenbegia albomarginata* **E**
FAMILY CAMPANULACEAE
Alpine and subalpine grasslands, herbfields and fellfields NI, SI and Stewart Is.
Small, perennial herb. Leaves green, somewhat lanceolate, in small flattened rosettes. Flower pale blue to pale violet to white, often with fine violet veins, on tall, slim, green or brownish-green stem to 250 mm, remains upright in flower. Seed a glossy brown capsule.

Similar species: Maori Bluebell *W. pygmaea* is smaller and found in NI and SI.

Korikori *Ranunculus insignis* **E**
KOPATA URAURA, HAIRY ALPINE BUTTERCUP, MOUNTAIN BUTTERCUP
FAMILY RANUNCULACEAE
Alpine grasslands, herbfields and rocky areas NI and SI, from East Cape to Kaikoura.
Attractive upright alpine herb to 900 mm. Leaves large, glossy green, circular with uneven edges, undersides densely hairy. Flowers large, golden yellow on upright, branching stem.

Similar species: Godley's Buttercup *R. godleyanus* has heavily veined leaves on long petioles and occurs from Lewis Pass to Mt. Cook Aoraki.

Snow Buttercup *Ranunculus nivicola* **E**
MOUNTAIN BUTTERCUP
FAMILY RANUNCULACEAE
Alpine and subalpine grasslands, fellfields and scree, to snowline on Mt. Taranaki.
Medium-sized, to 300 mm, very visible plant. Leaves large, almost circular, glossy green and heavily indented or lobed. Flowers large and bright yellow.

Vegetable Sheep [flower]

Maori Onion   Korikori

Rimuroa

Snow Buttercup   New Zealand Harebell

NATIVE HERBS – ALPINE AND SUBALPINE, COLOURED FLOWERS   **211**

ALPINE DAISIES

New Zealand boasts a wide variety of mountain daisies (family Asteraceae), with over 60 species of the genus *Celmisia* alone. All of these, and most of the other species of other genera, have white daisy-like flowers with yellow centres and fluffy wind-borne seeds.

Large Mountain Daisy *Celmisia semicordata* **E**
TIKUMU, SILVERY COTTON PLANT
Widespread in SI alpine and subalpine grasslands, herbfields and fellfields.
Large, to 600 mm, tufted perennial herb. Leaves long and broadly lanceolate, green with covering of white down when young, darker green when older. Flowers large on single bearded stem. Seedhead hemispherical. Largest of the alpine daisies.

Similar species: **Marlborough** or **Monro's Daisy** *C. monroi* is similar in appearance but smaller, with a very bearded flower stem. It is found only in Marlborough.

Dainty Daisy *Celmisia gracilenta* **E**
PEKAPEKA, COMMON MOUNTAIN DAISY
Widespread montane to alpine grasslands, herbfields and shrublands NI, SI and Stewart Is.
Slender, tufted daisy, to 250 mm. Leaves dark green, very narrow with margins rolled under, silvery underneath. Flowers single with spread out petals, on long, woolly, slightly bearded stems.

Lance-leaved Daisy *Celmisia armstrongii* **E**
SI alpine grasslands and herbfields.
Large, tufted daisy. Leaves green, thick, leathery, long (to 350 mm), narrow, longitudinally ridged, pointed. Margins revolute, undersides downy apart from prominent midrib. Flowers singly on erect somewhat bearded stems, to 300 mm.

Similar species: **Fiordland Mountain Daisy** *C. holocericea* very similar but midrib yellowish.

Cotton Daisy *Celmisia spectabilis* **E**
TIKUMU, PUAKAITO, PUHERETAIKO, COTTON PLANT
NI and SI alpine and subalpine grasslands, herbfields and fellfields Coromandel southwards.
Stout, tufted daisy. Leaves to 150 mm, dark green, pointed, elliptic, somewhat glossy above, white and downy underneath, in low rosettes sometimes forming large patches. Flowers on stout woolly stems to 250 mm.

Brown Mountain Daisy *Celmisia traversii* **E**
Alpine grasslands and herbfields, wetter areas, SI only.
Large, distinctive, tufted daisy. Leaves to 250 mm on long petioles, tongue-shaped and pointed, dark green above, margins and undersides rich brown. Flowers on tall brown bearded stems to 300 mm.

Celmisia densiflora **E**
SI alpine and subalpine grasslands, herbfields and fellfields, drier areas preferred.
Large tufted daisy. Leaves, to 150 mm, lance-shaped, pointed, toothed, teeth almost spine-like, margins sometimes wavy, underside downy. Flowers white, on tall greyish stems to 300 mm with leafy bracts.

Haast's Daisy *Celmisia haastii* **E**
SI alpine grasslands and herbfields to snowline.
Stout, spreading, patch-forming daisy. Leaves to 80 mm, elliptic, pointed, dark green above with distinct longitudinal grooves, silvery and downy underneath. Flowers singly on bearded stem to 200 mm. Plant may be sticky to the touch.

Mount Egmont Daisy *Celmisia major* var. *brevis* **E**
NI only. Herbfields and fellfields to 1400 m, on Mt Taranaki.
Short, to 150 mm, tufted daisy. Leaves narrow, pointed, rather leathery, dark green with distinct midrib and recurved margins, underside downy. Flowers white on short, greyish stem.

Similar species: *C.m.* var. *major* is similar but is found in lowland and coastal areas.

Dainty Daisy   Lance-leaved Daisy   Large Mountain Daisy

Haast's Daisy   *Celmisia densiflora*

Cotton Daisy   Brown Mountain Daisy   Mount Egmont Daisy

Larch-leaved Daisy *Celmisia laricifolia* **E**
Alpine regions in grasslands and rocky fellfields NI, SI and Stewart Is.
Small, to 100 mm, tufted daisy. Leaves, to 20 mm, dark green with silver down, in small rosettes, narrow and pointed with margins rolled under. Dead leaves often much in evidence. Flower white with lemon-yellow centre on thin stem, one per rosette.

Everlasting Daisy *Anaphalioides bellidioides* **E**
STRAWFLOWER, HELLS BELLS
Widespread in montane and subalpine areas, most habitats from Coromandel southwards, including Stewart Is.
Low spreading perennial daisy. Leaves in small flattened rosettes, green above, sometimes hairy, white and downy underneath. Flowers in profusion on short, to 100 mm, woolly stems.

Dwarf Alpine Daisy *Brachyglottis bellidioides* **E**
Widespread in SI alpine and subalpine grasslands, herbfields and fellfields.
Small, yellow-flowered daisy. Leaves green above, downy underneath, ovoid with long petioles, clearly indented midrib and network of veins. Flower yellow, singly on tall, 100–300 mm, very thin, slightly bearded stem.

Similar species: Two subspecies *bellidioides* and *crassa*, both restricted to SI.

South Island Edelweiss *Leucogenes grandiceps* **E**
Widespread in alpine and subalpine fellfields and rocky areas SI and Stewart Is.
Small, spreading, scrambling daisy. Leaves grey-green to near purple, on short, to 100 mm, upward-curving branches like overlapping rosettes. Flowers small and yellow in clusters surrounded by a ring of white woolly petal-like, bracts.

Similar species: Three other species are found, **North Island** *L. lentopodium*; **Marlborough** *L. neglecta*, and **Mount Peel** *L. tarahaoa.* All identified by location.

Woollyhead *Craspedia uniflora* **E**
Alpine and subalpine grasslands, herbfields, fellfields, scrublands and rocky areas NI, SI and Stewart Is.
Small to medium-sized tufted daisy. Leaves light to mid-green, pointed, in basal rosette as well as on flower stem. Flower on tall, 100–450 mm, stems with leaf-like bracts, white, or occasionally yellowish, hemispherical, developing into spherical fluffy seedheads.

Similar species: *C. lanata* is similar to *C. uniflora* but with grey-green rather downy leaves; while *C. incana*, is larger with near-white, very downy leaves and a yellow flowerhead.

Smooth Mat Daisy *Raoulia glabra* **EG**
Montane to alpine rocky areas and riverbeds, from Taupo in NI southwards, and east of the Main Divide in SI.
Low, to 20 mm, mat-forming daisy. Leaves light green, hairless, very short and pointed, in tightly packed rosettes. Flowers white, almost stemless, one per rosette.

Large-flowered Mat Daisy *Raoulia grandiflora* **EG**
NI and SI alpine and subalpine grasslands, herbfields, fellfields and rocky areas from Coromandel southwards.
Small mat-forming daisy. Leaves very short, purplish, in small tightly packed rosettes. Flowers white, stemless, often as large as leaf rosettes, one per rosette.

Silver Cushion Daisy *Celmisia argentea* **E**
Alpine and subalpine grasslands, herbfields and fellfields, especially damp areas SI and Stewart Is.
Low, to 20 mm, spreading, tufted daisy. Leaves rigid, short, pointed and silver-green in small rosettes. Flowers yellow with white petal like bracts, appearing within the leaves, one per rosette.

Larch-leaved Daisy

Everlasting Daisy

South Island Edelweiss

Dwarf Alpine Daisy

Woollyhead

Smooth Mat Daisy

Silver Cushion Daisy

Large-flowered Mat Daisy

NATIVE HERBS – TUFTED OR GRASS-LIKE PLANTS

Kakaha *Astelia fragrans* **E**
BUSH FLAX, BUSH LILY
FAMILY ASTELIACEAE
Widespread lowland to subalpine forest NI, SI and Stewart Is.
Large, to 1.2 m, tussock-like forest floor plant. Leaves long, to 1.5 m, broad and drooping with prominent vein either side of midrib. Inflorescence large, branched, flowers greenish white, sweet-scented, in spring. Fruit is an orange berry in yellow cup. Fruit may take 12 months to ripen.

Similar species: **Shore Astelia** *A. banksii* has silvery grey-green leaves and pink fruit, mainly coastal, enjoys dry conditions. **Kauri Grass** *A. trinervia* is smaller, the leaves are tightly folded at the base and have three veins either side of the midrib. Flowers are maroon and produced in the autumn, fruit crimson.

Mountain Astelia *Astelia nervosa* **E**
FAMILY ASTELIACEAE
Widespread in damp alpine and subalpine grasslands, herbfields and fellfields NI, SI and Stewart Is.
Distinctive, large, tufted grass-like plant to 750 mm. Leaves grey-green, long, to 1.5 m, narrow, pointed, hairy and drooping. Flowers greenish-yellow, sweet-scented, on large branched spike partly hidden within leaves. Fruit orange-red berry on female plant only.

Harakeke *Phormium tenax* **E**
FLAX
FAMILY HEMEROCALLIDACEAE
Widespread in coastal areas, lowland to montane forests, open areas, streamsides NI, SI and Stewart Is.
Very large unmistakable tufted plant, often forming impenetrable thickets. Leaves long, to 2.5 m, dark green, broad, often drooping. Flower spike very tall, to 3.5 m, branching with clusters of upturned trumpet-shaped red flowers.

Mountain Flax *Phormium cookianum* subsp. *cookianum* **E**
WHARARIKI
FAMILY HEMEROCALLIDACEAE
Alpine and subalpine grassland and scrub NI, SI and Stewart Is.
Large tufted grass-like plant. Leaves long, to 1 m, olive green, narrow, pointed, erect. Flowers red or orange-red, in lateral clusters on tall, to 2 m, dark brown spike. Fruit glossy black in pendulous pods.

Similar species: *P. cookianum hookeri* has longer, more drooping leaves, greenish flowers, very distinctive long drooping seed pods or capsules. It is found in coastal to subalpine habitats, especially cliffs and rocky areas.

Mikoikoi *Libertia ixioides* **E**
NEW ZEALAND IRIS, TUKAUKI, MANGA-A-HIRIPAPA
FAMILY IRIDACEAE
Coastal and lowland forest, stream banks and forest openings NI, SI and Stewart Is.
Tufted forest floor plant to 400 mm. Leaves light to yellow-green, erect with orange midrib. Flower delicate, white, three-petalled. Fruit bright yellow, becoming brown, held upright.

Turutu *Dianella nigra* **E**
INKBERRY, NZ BLUEBERRY
FAMILY HEMEROCALLIDACEAE
Lowland and lower montane forest margins and trails NI, SI and Stewart Is.
Tufted plant to 600 mm, sometimes forming large stands. Leaves midgreen, distinct midrib giving M-shaped cross-section. Flowers white with prominent yellow stamens on tall, thin, branched stem. Fruit whitish becoming deep blue-black.

Mountain Flax   Harakeke

Turutu [fruit]   Harakeke flowers

Mikoikoi [flower]   Kakaha [flower]   Mountain Astelia [flower]

NATIVE HERBS – TUFTED OR GRASS-LIKE PLANTS   **217**

NATIVE GRASSES, SEDGES AND RUSHES

New Zealand has a modest 190 native species of grass, 15 native rushes and 43 native sedges, plus large numbers of introduced species which dominate the agricultural and urban areas. While they are sorted by habitat here, there is considerable overlap. (**E** after the botanical name = endemic species and **EG** = endemic genus. **N** = native species.)

COAST AND SHORE

Blue Shore Tussock *Poa astonii* **E**
FAMILY POACEAE
Coastal cliffs, banks, shrublands, scrubs and coastal montane to 400 m NI, SI and Stewart Is.
Small, 150–250 mm, compact, tufted grass. Leaves narrow, rolled, green to blue-green, drooping. Inflorescence compact, straw-coloured, drooping.

Sand Tussock *Austrofestuca littoralis* **N**
HINAREPE, POUAKA
FAMILY POACEAE
Coastal areas, especially sand dunes, NI and SI.
Distinctive yellow-green tussock grass with thin, upright, to 700 mm, leaves. Dead leaves straw-coloured. Seedhead does not overtop leaves and has a zigzag shape when seeds fall.

Sand Sedge *Carex pumila* **N**
FAMILY CYPERACEAE
Coastal sandy areas, occasionally inland NI, SI and Stewart Is.
Rather tufted, to 400 mm, grass with coarse blue-green, curved leaves, creeping rhizome and upright flower spike with 3–7 stout, rather knobbly, spikes close together.

Pingao *Ficinia spiralis* **EG**
GOLDEN SAND SEDGE, PIKAO
FAMILY CYPERACEAE
Coastal sand dunes NI, SI and Stewart Is. An important stabilizing plant.
Low, spreading, grass-like plant with long surface rhizomes. Leaves long, to 300 mm, pointed, curving backwards, yellowish to golden green. Flower spike – dark brown seeds spirally arranged, like small bottle brush. An important dune stabilizer.

Spinifex *Spinifex sericeus* **N**
KOWHANGATARA
FAMILY POACEAE
Restricted to coastal sand dunes NI, SI and Stewart Is.
Distinctive, spreading dune grass. Leaves grey-green, to 400 mm, narrow, in-rolled. Male inflorescence at end of spike, several bushy spikelets. Female inflorescence resembles spiky ball; breaks off when ripe and blows along beach. Sends out long, creeping rhizomes to anchor sand. Spinifex and Pingao leave considerable open spaces on the dune which is important for other plant species, invertebrates and reptiles; while introduced **Marram Grass** *Ammophila arenaria* does not and often excludes Spinifex, Pingao and other beach plants. Marram is also a more efficient sand stabiliser which can lead to alteration of the coast.

Coastal Cutty Grass *Cyperus ustulatus* **E**
GIANT UMBRELLA SEDGE
FAMILY CYPERACEAE
Coastal and lowland open areas, especially wetlands and margins NI and SI, not found south of Christchurch.
Large, tufted sedge. Leaves long 0.6–2 m, and narrow, mid to yellow-green, with prominent midrib, often somewhat drooping. Inflorescence a large dark brown spray at tip of spike with rosette of stiff leaves. Individual flower rosettes elongated, spiky.
　　Similar species: Two introduced species are similar but have round inflorescences, **Umbrella Sedge** *Cyperus eragrostis* from the Americas has green or light brown, rather feathery ones, while **Purple Cutty Grass** *Cyperus congestus*, an import from South Africa, has dark purplish ones. Both are more common in NI than SI.

Blue Shore Tussock

Sand Tussock

Pingao

Pingao rhizome and flower spikes

Spinifex

Spinifex, female inflorescences

Coastal Cutty Grass

Purple Cutty Grass

Sand Sedge

Coastal Woodrush *Luzula banksiana* **E**
FAMILY JUNCACEAE
Widespread coastal and lowland areas, sand dunes and shrublands NI, SI and
Stewart Is. Five subspecies with varied distribution.
Short, to 200 mm, tufted rush. Leaves broad, glossy green, sometimes red-tipped.
Inflorescence on short, stiffly erect stems, flowers in dense brown cluster.

Sea Rush *Juncus kraussii australiensis* **N**
FAMILY JUNCACEAE
Coastal sand and salt marshes and tidal estuary margins, south to Westland and Otago.
Occasionally inland and at geothermal events around Rotorua.
Medium-sized erect, to 1.2 m, rush, forming large and often extensive clumps. Stems thin,
to 3 mm, greenish brown, leaves green, 1–2 per stem, similar but shorter. Inflorescence
small, erect, near stem tip, irregularly branched, flowers on tips.

WATERSIDE AND WETLAND

Purei *Carex secta* **E**
MAKURA, NIGGERHEAD   FAMILY CYPERACEAE
Widespread in coastal to montane wetlands NI, SI and Stewart Is.
Large tussock-forming sedge, sometimes forming short trunk (to 1 m). Leaves long, to 1 m,
dark to orange-green, narrow, drooping. Inflorescence much-divided, dark brown, drooping.

Rautahi *Carex geminata* **E**
CUTTY GRASS   FAMILY CYPERACEAE
Widespread coastal to lower montane wetlands, stream banks, forest clearings NI, SI,
Stewart Is.
Medium-sized tufted sedge. Leaves, 0.3–1.2 m, bright green to yellow-green, especially
smaller plants, and M-shaped in cross section. Inflorescences long, bunched, drooping,

Wiwi *Juncus pallidus* **N**
GIANT RUSH FAMILY JUNCACEAE
Coastal and lowland, wetlands and saltmarshes, often 'weed' in pastures, NI, SI and Stewart Is.
Large, to 2 m, leafless rush, forming substantial clumps or patches. Stems dark green,
pith-filled. Flowers in dense brown inflorescence.

Edgar's Rush *Juncus edgariae* **N**
WIWI FAMILY JUNCACEAE
Very widespread coastal to montane, occasionally alpine, grasslands and shrublands NI, SI
and Stewart Is., often 'invades' pastures and urban areas.
Large, 0.6–2.5 m, leafless rush, forms dense or diffuse tussocks. Stems light to dark green,
pith-filled. Inflorescence brown tufts, smaller, more compact that *J. pallidus,* often at stem tips.

Slender Clubrush *Isolepis cernua* **N**
FAMILY CYPERACEAE
Coastal and lowland wetlands, lake margins, NI, SI and Stewart Is. Occasionally inland.
Small, to 100 mm, somewhat prostrate, tufted or spreading sedge. Leaves bright to
yellowish green. Inflorescence at or close to tip of stems, becomes dark brown or black on
maturity of nut.

Raupo *Typha orientalis* **N**
BULRUSH   FAMILY TYPHACEAE
Coastal, lowland wetlands and margins, occasionally ditches and industrial sites, NI and SI.
Large, distinctive, erect rush, to 3 m, often forms large patches. Leaves long, sword-shaped,
green. Inflorescence dark brown, furry 'cylinder' on top of erect spike, shorter than leaves.

Jointed Twig Rush *Baumea articulata* **N**
FAMILY CYPERACEAE
NI coastal and lowland wetlands and swamps, in water or on margins.
Large, upright, to 2 m, rush, often forming dense patches. Leaves thin, tubular, dark green,
pointed. Inflorescence a large drooping, many-branched, mid to dark brown, panicle to
300 mm long.

Coastal Woodrush

Rautahi Sea Rush

Purei Wiwi Slender Clubrush

Jointed Twig Rush Raupo Edgar's Rush

NATIVE GRASSES, SEDGES AND RUSHES – WATERSIDE AND WETLAND **221**

FOREST

Broad-leaved Sedge *Machaerina sinclairii* **E**
FAMILY CYPERACEAE
NI only, streamsides and cliffs, forest clearings, south to Rimutaka Range.
Large, to 1 m, spreading, tussock-like sedge with long, to 1.2 m, drooping dark-green leaves. Inflorescence large brown, dense, drooping panicle.

Giant Sedge *Gahnia xanthocarpa* **E**
MAPERE, TUPARI MAUNGA   FAMILY CYPERACEAE
NI, damp areas in coastal and montane forest up to 800 m.
Very large, to 5 m, tussock-forming sedge. Leaves dark green, heavily drooping, rough to the touch. Inflorescence long, heavily branched, mid-brown, drooping. Fruit bright yellow nut maturing black.

Similar species: **Mapere** *G. rigida* is found on NI and SI and is of similar size and appearance but has rigidly erect flowerheads or panicles. Fruit are always bicoloured.

Hunangamoho *Chionochloa conspicua* **E**
BROAD-LEAVED BUSH TUSSOCK   FAMILY POACEAE
Coastal areas, streamsides, open lowland forest and shrublands NI, SI and Stewart Is.
Large to 1 m, green tussock, with long, to 1.3 m, narrow, hairy, drooping leaves, flattened at base. Inflorescence cream or pinkish, diffuse and very prominent on 2 m flower spike.

Bush Rice Grass *Microlaena avenacea* **N**
OAT GRASS   FAMILY POACEAE
Widespread in coastal to montane forest NI, SI and Stewart Is., especially clearings and alongside trails. Commonly with Hook Sedge but broader and darker leaves distinguish it.
Distinctive forest grass forming small tussocks or patches. Leaves 300–600 mm, broad and dark, sometime blue-green. Inflorescence tall, to 1 m, erect, diffuse and very prominent.

Hookgrass *Uncinia uncinata* **N**
KAMU, MATAU A MAUI, BASTARD GRASS   FAMILY POACEAE
Widespread in coastal and lowland forest and scrub, especially alongside walking tracks, NI, SI and Stewart Is.
Medium-sized, leafy sedge forming small tussocks. Leaves to 500 mm, bright green, narrow and pointed. Inflorescence to 80 mm, on erect, to 800 mm, spike, yellow-green becoming dark brown as seeds mature. Seeds attach readily to passing bipeds and quadrupeds.

Forest Sedge *Carex dissita* **E**
FAMILY CYPERACEAE
Widespread from coast to montane forest and scrub NI, SI and Stewart Is., especially streamsides and other damp locations. Often with *C. solandri*.
Medium-sized sedge, leaves to 500 mm, dark green to yellow green, varying from broad to narrow; M-shaped cross section. Inflorescences quite short, spaced out along spike, male at tip, others female and oblong. Seed covers bicoloured.

Solander's Forest Sedge *Carex solandri* **E**
FAMILY CYPERACEAE
Coastal to montane forest, damp areas, streamsides NI, SI, Stewart Is., often with *C. dissita*.
Medium-sized sedge to 1 m grows in small clumps. Leaves, to 1.2 m, yellow-green, drooping, 2–6 mm wide. Flowerhead drooping on long slender peduncle, seed casing black.

Oplismenus hirtellus **N**
FAMILY POACEAE
Forest floor in coastal and lowland forest NI only, often along trails. Two varieties: var. *imbecillus* is widespread in NI; var. *hirtellus* Northland only.
A distinctive prostrate, to 100 mm, spreading grass. Leaves green, narrow, pointed, elliptic, and widely spaced. Inflorescence on fine, slightly drooping spike to 200 mm.

Broad-leaved Sedge

Bush Rice Grass

Giant Sedge [inflorescence]

Oplismenus hirtellus

Hookgrass [seedhead]

Hunangamoho

Forest Sedge

Solander's Forest Sedge

TUSSOCK GRASSES

There are over 30 species of tussock grass plus numerous subspecies, these are largely confined to drier and upland areas. Many are now used as ornamentals. The following species are all in the family Poaceae.

Hard Tussock *Festuca novae-zelandia* E
FESCUE TUSSOCK

Widespread in higher montane and subalpine grasslands and shrublands NI, SI and Stewart Is.
An erect, 300–600 mm, light brown grass, with fine, slightly rough leaves. Inflorescence diffuse, erect, prominent. Grows well spaced out in small tussocks.

Silver Tussock *Poa cita* E

Widespread in montane shrublands, grasslands and stony areas, especially damp locations NI, SI and Stewart Is.
Small erect, 300–700 mm, tussock with mixed green and light brown leaves giving it a silvery appearance. Inflorescence diffuse, erect, prominent. Grows spaced out in tussocks.

Red Tussock *Chionochloa rubra* E

Higher montane to alpine grasslands, occasionally shrublands NI, SI and Stewart Is.
Distinctive russet-brown tussock to 1 m with drooping leaves, often dominant and forming large dense stands. Leaves long, to 1.5 m, and very narrow. Inflorescence diffuse and barely topping leaves. Four subspecies: *rubra* var. *rubra* Volcanic Plateau to North Canterbury, though scarce in South; *rubra* var. *inermis* found only on Mt. Taranaki where it is dominant; *occulata* found in Nelson and the West Coast south to Mt. Aspiring NP; and *cuprea* or **Copper Tussock** found from North Canterbury south to Stewart Is.

Broad-leaved Snow Tussock *Chionochloa flavescens* E

Coastal to montane to alpine, shrublands, grasslands and cliffs NI, SI and Stewart Is.
A large, to 1.5 m, green tussock. Leaves long, to 2 m, drooping, without conspicuous midrib. Inflorescence large, dense and very prominent. Four subspecies: *brevis* SI only; *flavescens* NI and SI; *lupeola* NI and SI; *hirta* SI only, especially Mt. Cook NP.

Similar species: *C. flavicans* had two forms **C. f.** *flavicans* NI only, Coromandel to Hawke's Bay, coastal to subalpine cliffs and **C. f.** *temata* found Te Mata Peak, Hawke's Bay only; **Narrow-leaved Snow Tussock** *C. rigida* is browner and with much smaller, more sparse and less conspicuous inflorescences. It is found at higher altitudes on SI and Stewart Is. **Mid-ribbed Snow Tussock** *C. pallens* is significantly smaller, has broader leaves with pale midrib on the underside. It is found in alpine and subalpine grasslands.

Toetoe *Cortaderia* spp. E

Widespread coastal to subalpine, frequently alongside roads, tracks or streams NI, SI and Stewart Is. Smaller variety occurs in Northland only.
Very large tussock with long, to 2.5 m, green drooping leaves with midrib and two lateral ribs. Inflorescence large, graceful, drooping, white, sometimes pinkish, dense, and very prominent on a flower spike to 6 m. There are four endemic species, three NI and one SI, as well as one introduced species. The flower spikes of *C. fulvida* are up to 3.5 m long, widespread in NI, from coast to subalpine, especially damp areas and roadsides. *C. toetoe* is taller and less widely distributed, occurring from Tauranga south to Wellington, preferring wetlands, often with native flax *Phormium tenax*. *C. splendens* is the tallest species, reaching 6 m, but is largely coastal especially on dunes and cliffs. In SI, *C. richardii*, the smallest and most graceful toetoe, has flower spikes up to 3 m long.

All native species flower in the spring while the invasive introduced **Pampas Grass** *C. selloana* is autumn flowering, with very erect flowerheads. It also differs in that its leaves have a single prominent midrib and when dead curl up and gradually disintegrate like wood shavings while the leaves of the native species fold in half and are discarded intact.

Carpet Grass *Chionochloa australis* E

Upper montane to alpine areas from Nelson to Canterbury.
Distinctive, short, to 100 mm, mat-forming grass, often almost entirely exclusive. Flower spike much taller than leaves, to 400 mm.

Hard Tussock   Silver Tussock

Red Tussock   Red Tussock

Broad-leaved Snow Tussock

Carpet Grass   Toetoe   Pampas Grass

NATIVE GRASSES, SEDGES AND RUSHES – TUSSOCK GRASSES   **225**

FERNS AND THEIR ALLIES

With its mild climate, high rainfall and extensive bush, New Zealand is ideal habitat for ferns. There are over 165 native species, over 80% of which are endemic. They range from the 15-m tall Mamuka *Cyathea medullaris* to the lovely delicate filmy ferns. They are a key and fascinating feature of the New Zealand bush. Apart from tree ferns, I have classified them by the pinnate nature of their fronds. (**E** after the botanical name = endemic species, **EG** = endemic genus and **N** = native species.)

TREE FERNS

Tree ferns are a common, characteristic and attractive feature of the New Zealand bush. There are eight species from two genera, on the main islands with two more species in the Kermadecs. The young fronds of *Cyathea* are covered in scales, while *Dicksonia* has hairs. Tree ferns colonise open south-facing slopes and clearings in the forest. Their trunks are fibrous rather than woody and their ancestry dates back at least 100 million years.

SIMPLE TREE FERN GUIDE

Mamaku	*Cyathea medullaris*	Tallest tree fern; large black-stiped fronds which fall off leaving oval scars.
Ponga, Silver Fern	*C. dealbata*	New frond stipes greenish white, underside of fronds silver/white. Fronds fall off when dead, stem bases clearly visible as part of trunk.
Katote, Soft Tree Fern	*C. smithii*	Skirt formed from midribs of dead fronds.
Punui, Gully Tree Fern	*C. cunninghamii*	As Mamaku but more slender and dead frond bases remain black and clearly visible on trunk, less prominent than *C. dealbata*.
Creeping Tree Fern	*C. colensoi*	Small short trunk to 1 m, often prostrate, straw-coloured scales on frond stems. 3x pinnate.
Wheki	*Dicksonia squarrosa*	Fronds fall off leaving 'litter' of rust-brown fronds. Frond stumps remain visible on trunk. Thick brown/black hairs on stipes.
Wheki Ponga	*D. fibrosa*	Thick, often voluminous, skirt of dead fronds.
Woolly Tree Fern	*D. lanata*	Small to 2 m, trunk 100–120 mm. Fronds few in number, stipes slender with dense brown hairs at base. South of Whangarei, generally prostrate with branching trunk. 4x pinnate.

Note: Cyathea – scales; Dicksonia – hairs.

FAMILY CYATHEACEAE

Mamaku *Cyathea medullaris* **N**
BLACK TREE FERN, KORAU, PITAU
Widespread in lowland forest on all three main islands, mainly in the west on SI.
A large emergent tree fern (H to 20 m/TD to 300 mm), very large 3x pinnate fronds up to 6 m in length and 2 m wide, with heavy black stipes. The fronds persist on young plants, but are discarded on mature plants, leaving distinctive whitish, spikey, oval scars on the trunk.

Mamaku

Punui *Cyathea cunninghamii* **N**
GULLY OR SLENDER TREE FERN
Moist coastal forests south to northwest of SI.
A tall, emergent, slender and elegant tree fern (H to 20 m/TD to 120 mm). Trunk tends to taper towards base, fronds discarded with stipe bases melded onto trunk, leaving distinctive near-black lumpy trunk. Fronds 3x or 4x pinnate and up to 3 m long and 1.2 m wide. Not common, but found in small groups, mainly coastal.

Ponga *Cyathea dealbata* **E**
SILVER FERN, SILVER TREE FERN
Drier coastal and montane forest NI and SI.
Medium-sized tree fern (H to 10 m/TD to 200 mm) subcanopy species with fronds 3x pinnate and up to 4 m long and 1.2 m wide. Young stipes and underside of fronds are whitish giving it a silver appearance. Fronds break off leaving ragged stump giving the brown trunk a rough rasp-like appearance. One of New Zealand's national emblems.

Katote *Cyathea smithii* **E**
SMITH'S OR SOFT TREE FERN, WHE
Cooler, wetter forest to 600 m NI, lower in SI and Stewart Is.
Fairly short, subcanopy tree fern (H to 8 m/TD to 250 mm) with a rather flat crown. Fronds 3x pinnate and up to 2.5 m long and 750 mm wide. Distinctive skirt of dead frond stipes hanging beside trunk.

Creeping Tree Fern *Cyathea colensoi* **E**
MOUNTAIN TREE FERN
Moist montane forests from Bay of Plenty southwards, including Stewart Is.
Small, often prostrate tree fern (H to 1 m/TD to 80 mm), fronds 3x pinnate and up to 1.5 m long and 600 mm wide. Distinctive brown hairs on stipes, especially near bases, giving a furry appearance.

FAMILY DICKSONIACEAE

Wheki *Dicksonia squarrosa* **E**
ROUGH TREE FERN
Widespread on all three islands in coastal to montane forest, sometimes swamps.
Very common, medium-sized, slender subcanopy tree fern (H to 7 m/TD to 120 mm). Fronds are 3x or 4x pinnate and up to 2.5 m long and 1 m wide. High crown with slender, near-black frond stipes growing near vertical at first. Fronds discarded haphazardly, giving overall untidy appearance. Spreads by stolons, forming groves carpeted in discarded fronds.

Wheki Ponga *Dicksonia fibrosa* **E**
KURIPAKA
Found in coastal and montane forest from Auckland southwards.
Medium-sized, subcanopy, thick-trunked tree fern (H to 6m/TD to 600 mm). Produces large number of fronds, up to 30 at a time. Fronds are 3x or 4x pinnate and up to 2.4 m long and 600 mm wide. Very distinctive broad skirt of fronds complete with pinnae, which often obscures the very substantial trunk.

Tuakura *Dicksonia lanata* **E**
WOOLLY OR STUMPY TREE FERN
Lowland Kauri forest, lowland and montane forest, NI and SI.
A small, either erect or prostrate, branching or stoloniferous, subcanopy tree fern (H to 2 m/TD to 100 mm). Fronds either 3x or 4x pinnate up to 2 m long and 900 mm wide. Erect form found in Northland, but prostrate from Whangarei southwards. Slender stipes, densely hairy at base.

Ponga Katote

Creeping Tree Fern Punui Wheki Ponga

Wheki Wheki fronds Tuakura

TRUE FERNS – FERNS WITH UNDIVIDED FRONDS

Common Strap Fern *Grammitis billardierei* **N**
Common in lowland forest to subalpine bush NI, SI and Stewart Is.
A common epiphytic, rupestral or occasionally terrestrial fern with undivided blunt-tipped frond (20 mm long by 9 mm wide), stipe minimal. Rhizome is erect or short, creeping. The stipe is indistinct but winged. The sori are arranged in herringbone pattern.

Grammitis patagonica **N**
Beech forest, subalpine and alpine areas, Urewera Ranges south to Dunedin.
Similar to *G. billardieri* but with shorter, narrower fronds (150 mm long by 10 mm wide), tips generally more rounded. Stipe longer (to 50 mm) not winged and distinctly hairy. Sori more rounded and confined to upper end of frond. Rhizome long, creeping usually on rock.

Lance Fern *Anarthropteris lanceolata* **EG**
Lowland and montane forest NI and coastal areas of northern SI.
Bright green epiphytic or rupestral fern with long, pointed, slightly fleshy, undivided fronds (to 300 mm long by 20 mm wide). Midrib prominent but veins indistinct. Stipes short and winged. Sori round and creating bumps on upper surface of frond. Rhizome erect with creeping roots.

Kidney Fern *Trichomanes reniforme* **E**
KONEHU, KOPAKOPA, RAURENGA
Lowland to montane forests NI and west coast of SI.
Distinctive fern with bright green, shiny, near round or kidney-shaped frond (to 100 mm by 130 mm), stipe to 250 mm, generally smaller. Rhizome is creeping and thin stipes are erect. Plant tends to form mats on ground or rocks, occasionally epiphytic. Sori on frond margins.

Leather Leaf *Pyrrosia eleagnifolia* **E**
Widespread in coastal, lowland and montane areas, on trees and rocks, NI, SI, Stewart Is.
Distinctive epiphytic fern with pale grey-green, thick, fleshy frond (200 mm long by 20 mm wide), stipe slightly winged, to 30 mm, starts almost circular, becoming elongated with age. Sori circular pale brown.

Hymenophyllum lyallii **N**
Epiphytic or on rocks and banks, from Kaitaia southwards, including Stewart Is., west only.
Small, delicate, fan-shaped or near circular, fern. Laminae to 40 mm by 30 mm, stipe to 70 mm, branched hairs on stipe. Frond pale green, translucent, sometimes pinnate. Sori on tips of laminae.

Fan Fern *Schizaea dichotoma* **N**
Lowland forest from Kaitaia south to Taupo.
Small, inconspicuous terrestrial fern. Stipe green, erect to 300 mm, flattened and forked up to six times, forming a fan shape. Laminae are tiny (to 7 mm), pinnate and brownish at the tips of the stipe branches. Pinnae in 5–8 pairs, infolded up to 4 mm long. Sori on undersides of laminae.

FLOATING FERNS

Pacific Azolla *Azolla filiculoides* **I**
KARERARERA, RETORETO, RETURETU
Slow-moving streams, lakes and ponds from Dargaville southwards, eastern districts of SI.
Unmistakeable floating fern which forms dense red mats often covering entire surface of lake or pond. Fronds ovate or elliptical, branching (to 30 mm by 30 mm). Roots are simple and hang down into water. Reproduces mainly by frond division.

Common Strap Fern

Kidney Fern

Lance Fern

Grammitis patagonica

Hymenophyllum lyallii

Leather Leaf

Fan Fern

Pacific Azolla

TRUE FERNS – UNDIVIDED FRONDS   231

TRUE FERNS – FERNS WITH FRONDS ONCE PINNATE

Mangemange   *Lygodium articulatum*   **E**
MAKAMAKA, BUSHMAN'S MATTRESS
Lowland forest northern half of NI.
Unusual and distinctive vine-like fern. The climbing 'stem' is actually the rachis or midrib, of the frond which keeps on growing. Sterile pinnae, up to 100 mm long and twice forked. Fertile pinnae fork many times producing fan shape, with sporangia on spikes on end segments. Often in dense tangles and in tree tops. Used by Maori for ropes and twine.

Button Fern   *Pellaea rotundifolia*   **N**
TARAWERA, ROUND-LEAVED FERN
Dry rocky areas, lowland to montane bush and open areas south to Dunedin.
Distinctive dark green fern with creeping rhizome and dark brown scaly stipe to 200 mm, and rachis. Frond dark green and narrow to 600 mm. Laminae narrow and elongated. Pinnae near oblong with rounded ends. Sori around margins of pinnae apart from tips.

Tangle Fern   *Gleichenia dicarpa*   **N**
SPIDER FERN, SWAMP UMBRELLA FERN
Open scrubland, lowland to subalpine areas on main islands apart from east coast.
Unusual and distinctive fern with creeping rhizome. Frond upright to 1 m; stipe to 500 mm. Laminae are 3x or 4x branched and spread out horizontally from smooth, erect, brown stipe, giving a tangled interwoven appearance. Pinnae to 50 mm, are curled downwards, with one or two sporangia on each. **Carrier Tangle Fern** *G. microphylla* is similar but with up to three tiers of fronds all branching 3x to 4x and scrambling up to 2 m in height. Tips of pinnae flattened.

Umbrella Fern   *Sticherus cunninghamii*   **E**
TAPUWAE KOTUKU, WAEKURA
Open lowland and montane forest and roadsides, mainly west coast on SI, but also NI and Stewart Is.
Distinctive bright green fern with creeping rhizome. Frond broad and spreading (to 1 m) with stiff vertical stipe (to 500 mm), long drooping once pinnate laminae, may grow several tiers of laminae giving an umbrella-like appearance. Sori are yellow-brown and in rows either side of midribs.

Similar species: *S. flabellatus* is shorter, darker green, laminae more branched giving a fan-like appearance. Found only in Northland, Waikato and northwest of SI.

Kowaowao   *Microsorum pustulatum*   **N**
HOUND'S TONGUE, PARAHARAHA
Coastal to montane forests and open areas on NI, SI and Stewart Is., prefers drier habitats.
Bright glossy green fern with distinctive, large, 7–8 mm, greenish-grey creeping or climbing rhizome with dark scales. Stipe dark brown (to 250 mm), rachis and midrib brown. Young fronds strap-shaped and pointed, mature fronds pinnate, to 700 mm, with long, tongue-like ultimate pinna. Large round sori on either side of midrib.

Similar species: *M. novae-zelandiae*, largely epiphytic, longer fronds, much finer pinnae, rhizome covered in orange-brown hair-like scales. **Mokimoki** *M. scandens* is epiphytic and a vigorous climber on trees. Narrower and very variable fronds, juveniles undivided, thin rhizome with dark black-brown scales, sori much smaller and near margins. Prefers damper habitats.

Necklace Fern   *Asplenium flabellifolium*   **N**
BUTTERFLY FERN, WALKING FERN
Coastal to montane areas, mainly Hawke's Bay to South Canterbury.
Distinctive creeper-like fern with tufted rhizome and large number of narrow pale green fronds to 450 mm. Stipe, to 150 mm, rachis green. Pinnae are fan-shaped, with serrated margins and well separated. May form large patches with tips of fronds rooting and forming new plant. Sori to 6 mm long radiating from base of pinnae.

Mangemange

Button Fern

Umbrella Fern

Tangle Fern

Necklace Fern   Kowaowao

TRUE FERNS – ONCE PINNATE   **233**

Shining Spleenwort *Asplenium oblongifolium* **E**
HURUHURUWHENUA, PARENAKO, PARETAO
Coastal to lower montane areas in NI, mainly coastal areas in northern half of SI.
A common, upright, terrestrial fern. Dark green frond, to 1.5 m, whose erect rhizome forms a heavy woody base. Stipes, to 500 mm, are dark brown, rachis green but becoming brown towards base with age. The pinnae are bright glossy green and pointed. Sori round to 3 mm, away from margins.

Shore Spleenwort *Asplenium obtusatum* **N**
PARANAKO, PARENAKO, PARETAO
Coastal areas NI, SI and Stewart Is.
Similar to *A. oblongifolium* but dull green, slightly fleshy laminae, rhizome erect and woody, fronds to 850 mm. Stipe to 400 mm and base of rachis brown, green higher up. Pinnae blunt with serrated margins. Often stunted due to exposed location. Sori round to 1 mm but away from margins.

Subspecies *A.o. obtusatum* is found from Wellington southwards and *A.o. northlandicum* on northern coasts on NI, from Bay of Plenty to Taranaki.

Petako *Asplenium polydon* **N**
PERETAO, SICKLE SPLEENWORT
Coastal to lower montane areas in NI, mainly coastal areas in northern half of SI.
Attractive epiphytic fern with dark green, doubly toothed, pointed, triangular pinnae. Rhizome while creeping forms distinct tufts. Stipe to 650 mm, rachis dark brown and scaly. Fronds are long, to 1.85 m, drooping and tapering from base. Sori to 20 mm long, curving.

Asplenium scleroprium **E**
Southern coastal areas on SI and Stewart Is.
Attractive coastal fern with stout, erect rhizome. Frond upright, to 1 m; stipe to 500 mm with long, elliptical, dark green laminae. Pinnae are deeply and regularly toothed. Stipe and rachis are scaly, dark underneath, green on top. Sori long, to 10 mm, reaching to margins at indentations.

Nini *Blechnum chambersii* **N**
LANCE FERN, RERETI
Lowland and montane forests, especially stream banks, NI, SI and Stewart Is.
Common moisture-loving fern with stout, erect rhizome; frond narrow and to 600 mm long; stipe to 100 mm, hairy at base. Sterile lamina reddish becoming dark green, generally lance-shaped and with a slight lateral curve. Pinnae short, to 30 mm, slightly serrated margins, bluntly pointed, becoming shorter near base. Fertile frond shorter, with narrow, widely spaced pinnae completely covered in sori.

Peretao *Blechnum colensoi* **E**
COLENSO'S HARDFERN, PETAKO
Wet lowland to montane forests, by streams and waterfalls, NI, SI and Stewart Is.
Rather distinctive almost hand-shaped fern with creeping rhizome, with both sterile and fertile fronds to 600 mm. Stipe to 250 mm, dark brown, scaly at base, rachis green. Sterile fronds, strap-shaped or pinnate with several broadly elliptical rather leathery pointed pinnae, the ultimate being longest and in line with rachis. Fertile frond has finer pinnae.

Crown Fern *Blechnum discolor* **E**
PETIPETI, PIUPIU
Drier lowland to montane forest, especially beech, NI, SI and Stewart Is.
Handsome and common fern, often with short trunk up to 1 m, colonial, rhizome stout, woody, stoloniferous. Stipes brown and scaly at base, to 200 mm, rachis light brown. Fronds narrow and to 1.2 m long, many forming distinctive vase-shaped crown. Pinnae oblong, pointed with scalloped margins. Fertile fronds brown, slightly longer than sterile, with broad sterile pinnae near base and narrow fertile ones higher up. New fronds produced simultaneously.

Shining Spleenwort

Shore Spleenwort

Petako

Nini

Asplenium scleroprium

Peretao

Crown Fern

Climbing Hardfern *Blechnum filiforme* **E**
THREAD FERN
Coastal and lowland forest south to northern tip of SI.
An unusual creeping and climbing fern with two distinct phases and long creeping/climbing rhizome. Juvenile is terrestrial creeping with short stipes to 60 mm and small fronds up to 300 mm long. Adult phase epiphytic. Stipes to 100 mm and rachis brown and scaly. Fronds long, to 700 mm, gently drooping. Pinnae long, to 90 mm, pointed with toothed and somewhat scalloped margins. Fertile fronds have very long thread-like pinnae.

Creek Fern *Blechnum fluviatile* **N**
KIWAKIWA, KIWIKIWI
Damp areas in lowland and montane forests, especially by streams, NI, SI and Stewart Is.
Common dark green terrestrial fern with stout, erect rhizome, frond narrow and dark green to 850 mm). Stipe up to 100 mm long and rachis dark brown and densely scaly. Pinnae adnate, rounded or oblong, ends may be toothed. Fertile fronds slightly longer than sterile, pinnae short, narrow and held close to rachis.

Mountain Kiokio *Blechnum montanum* **E**
Montane and subalpine forest Coromandel and Pirongia southwards, excluding Stewart Is.
Large, often colourful, terrestrial fern with long, creeping rhizome; frond broad, to 700 mm long; stipe to 250 mm. Rachis dark brown with lighter scales. Pinnae long, to 140 mm, bluntly pointed with distinct midrib. Ultimate pinna longer than rest. Fertile frond similar but with very narrow upright pinnae. Replaces *B. novae-zelandiae* at higher altitudes.

Kiokio *Blechnum novaezelandiae* **E**
HOROKIO, PALM LEAF FERN
Lowland and montane regions on road banks and gully sides on all three islands.
Large, distinctive, graceful fern, with short creeping rhizome and very large drooping frond to 3.2 m; stipe to 700 mm and rachis dark with small light brown scales. Young lamina may have pinkish tinge. Pinnae adnate, long, to 350 mm, and finely tapering, scalloped on margins. Fertile fronds similar length, pinnae long and narrow, overlapping in mid-frond.

Alpine Hardfern *Blechnum penna-marina* **N**
LITTLE HARDFERN
Lowland to alpine areas on all three main islands, but uncommon north of Bay of Plenty.
Small, narrow-fronded terrestrial fern that grows in large often quite colourful patches with creeping rhizome and small narrow frond to 420 mm; stipe to 170 mm and rachis brown. Pinnae small, to 12 mm, adnate, rounded, pinkish bronze towards tip. Fertile fronds significantly longer than sterile with narrow, widely spaced pinnae.

Small Kiokio *Blechnum procerum* **E**
Lowland to subalpine areas, often forming large patches, NI, SI and Stewart Is.
Medium-sized terrestrial fern with long almost oblong pinnae and short, stout, creeping rhizome. Long broad frond to 950 mm, stipe to 400 mm, rachis brown with dark scales near base. Midrib prominent in pinnae, ultimate pinna longer than rest. Fertile frond may be longer than sterile, with narrow, well-spaced, upward-pointing pinnae.

Mountain Hardfern *Blechnum vulcanicum* **N**
KOROKIO
All three main islands montane forest, roadbanks, forest margins or coastal areas in far south.
Medium-sized terrestrial or rupestrial, often colonial, fern with erect or short creeping rhizome. Broadly tapering frond to 750 mm. Stipe to 400 mm, brown, rough and scaly at base, rachis brown. Lamina elliptic to triangular. Pinnae adnate, oblong with pointed apices, margins slightly toothed, midrib and veins distinct.

Pukupuku *Doodia australis* **N**
RASP FERN
Coastal and lowland areas, south to north coast of SI.
A rough-textured terrestrial fern growing in colourful patches with erect or short creeping rhizome. Frond to 600 mm. Stipe, to 250 mm, brown and hairy with scales. Rachis brown underneath, green above, slightly hairy. Lamina narrowly elliptic. Pinnae long, to 100 mm, adnate, stalked near stipe, yellowish or pinkish-red near tip. Indusia linear and hairy.

Climbing Hardfern　Creek Fern　Mountain Kiokio

Small Kiokio　Kiokio [growing tip]

Alpine Hardfern

Pukupuku　Mountain Hardfern

TRUE FERNS – FERNS WITH FRONDS MORE THAN ONCE PINNATE

ONCE OR TWICE PINNATE

Veined Bristle Fern *Trichomanes venosum* **N**
VEINED FILMY FERN
Wet coastal to montane forests on NI, SI, Stewart Is.

A delicate epiphytic fern with long, widely separated, finger-like, translucent pinnae. Rhizome long-creeping; frond to 200 mm. Stipe short, to 50 mm, rachis winged, lamina irregular elliptic, pinnae are rectangular and have distinct branching veins, the lower ones may be lobed or bipinnate. Sori solitary and held in indusia on upper margins of pinnae.

Lyall's Spleenwort *Asplenium lyallii* **E**
Coast to subalpine SI, Stewart Is. Prefers drier areas.

Attractive and distinctive shiny dark green terrestrial fern with stout, erect rhizome; frond to 600 mm; stipe, to 200 mm, brown, slightly scaly; rachis brown towards base, Lamina shiny, elliptical; pinnae narrow, triangular, often scaly, young ones toothed, mature ones becoming bipinnate. Sori up to 10 mm arranged in herringbone pattern.

TWICE PINNATE

King Fern *Marattia salicina* **N**
PARA, HORSESHOE FERN
Northland south to Coromandel and Urewera Ranges, also Taranaki.

Very large, elegant, dark green fern with erect, woody rhizome; frond to 4 m; stipe to 1 m. Lamina ovate with stalk of primary pinnae enlarged at base, secondary pinnae up to 200 mm long and oblong. Sori bean-shaped on margins of secondary pinnae. Rootstock is edible; population seriously affected by feral pig and other alien species.

Small Maidenhair *Adiantum diaphanum* **N**
Shaded banks and overhangs in coastal forests south to Nelson and Marlborough.

Small pale green fern with short trunk and erect rhizome. Frond, to 250 mm, one, occasionally two branches, giving a three-toed appearance, lamina narrowly triangular. Frond to 250 mm; stipe to 150 mm, and rachis dark brown, fronds delicate and somewhat transparent. Sporangia kidney-shaped, hairy indusia on outer margins of pinnae.

Hymenophyllum rarum **N**
On trees, rocks or banks lowland to montane forest, also in the open, NI SI, Stewart Is.

A pale green, delicate, translucent epiphytic fern with long, thin, creeping rhizome; lamina narrowly elliptical; frond to 200 mm. Stipe to 70 mm, very thin, rachis narrowly winged, lamina ovate with rather square-ended secondary pinnae. Sori at the ends of ultimate segments. Strong resistance to drought.

Gully Fern *Pneumatoteris pennigera* **N**
PAKAUROHAROHA, PIUPIU, FEATHER FERN
Damp coastal, lowland and montane forest, especially gulleys and streams NI, SI.

Large, dark green fern with short, slender trunk, to 1 m, and long, upright fronds to 1.75 m, drooping at the tips; lamina narrowly elliptic; stipes, to 250 mm, brown and scaly; rachis brown. Primary pinnae long, to 200 mm, secondary pinnae round-ended. Sori round either side of secondary pinnae midrib.

Drooping Spleenwort *Asplenium flaccidum* **N**
MAKAWE, RAUKATARI, HANGING SPLEENWORT
Widespread, coastal, lowland and montane forests, pine plantations on NI, SI and Stewart Is.

Distinctive terrestrial or epiphytic 'weeping' fern with short, erect rhizome; frond to 1.25 m; stipe, to 250 mm, brown at base, rachis green. Lamina elliptic, primary pinnae long, to 200 mm and narrow, secondary pinnae short and pointed. Sori to 10 mm long, close to margins.

Similar species: *A. haurakiense* found from Three Kings to Bay of Plenty has extended first secondary pinnae on each primary pinna – becoming partially tripinnate.

Veined Bristle Fern

Lyall's Spleenwort

Small Maidenhair

King Fern [secondary pinnae]

Hymenophyllum rarum

Gully Fern

Drooping Spleenwort

Polystichum silvaticum E
Lowland to montane forests, Auckland south to northern SI. Prefers damp, dark areas.
Distinctive, dark green terrestrial fern with very prickly fronds. Rhizome erect, scaly; frond
to 650 mm; stipe to 250 mm and rachis thickly covered with brown scales. Lamina elliptic,
secondary pinnae deeply divided and with holly-like spikes on apices. Sori round, grouped
on secondary pinnae. Smaller size, slight wings on pinna midrib and more open and
prickly appearance distinguish it from *P. vestitum* where ranges overlap.

Prickly Shield Fern *Polystichum vestitum* E
PUNIU
Common in lowland and montane areas, tussock grasslands, NI, SI, Stewart Is., higher
altitudes in north.
Large, handsome, dark green terrestrial fern with large erect rhizome, which builds to a
short trunk; frond to 1.5 m; stipe to 500 mm and rachis densely covered in brown scales.
Lamina narrowly elliptic, primary pinnae to 150 mm long, secondary pinnae deeply
dissected on larger fronds, and distinctly pointed.

Blechnum fraseri N
North Cape south to Tauranga, northwest Nelson south to Westport.
Its thin trunk, to 1.5 m, makes this fern look like a skinny miniature tree fern; often in small
colonies. Frond to 600 mm; stipe to 200 mm, scaly at base only, variable wing on upper
stipe and rachis. Lamina elliptic, primary pinnae to 200 mm long, pointed. Secondary
pinnae narrow and bluntly pointed. Fertile frond similar but thinner.

Cystopteris tasmanica N
Montane and subalpine areas and grasslands, on rock especially limestone, NI and SI.
Small drooping fern with long, to 180 mm, stipe; frond to 250 mm. Lamina ovate to oblong,
primary pinnae well spaced, secondary pinnae rounded and indented at tip. Indusia
pear-shaped around margins.

Royal Fern *Osmunda regalis* I
Wet and swampy areas, northern NI.
Large, imposing, pale green, very upright fern with erect rhizome sometimes forming short
trunk. Fronds to 2 m; stipes and frond yellow-green. Secondary pinnae rectangular with
rounded tips. Fertile fronds have sterile pinnae lower down and fertile ones at tips looking
like upright catkins.

TWICE OR THREE TIMES PINNATE

Common Maidenhair *Adiantum cunninghamii* E
On cliffs, banks and amongst rocks, in coastal and lowland forest, NI, SI, Stewart Is.
Medium-sized, delicate terrestrial fern with hairless branching fronds and creeping
rhizome. Frond to 550 mm; stipe, to 250 mm, dark and erect, rachis dark. Lamina elliptic
with ultimate segments almost oblong. Sporangia held in kidney-shaped indusia along
upper margins of pinnae.

Fan-like Filmy Fern *Hymenophyllum flabellatum* N
Widespread wetter lowland to montane forest, especially on tree fern trunks NI, SI, Stewart Is.
A delicate, epiphytic, translucent filmy fern with fan-shaped fronds, and a long, thin,
creeping rhizome. Frond to 370 mm; stipe, to 120 mm, thin with yellowish hairs near base.
Rachis narrowly-winged with a few yellow hairs. Lamina ovate to narrowly ovate, pinnae
fresh green, clearly veined, with smooth margins. Sori covered by indusia on ends of
segments.

Ground Spleenwort *Asplenium appendiculatum*
Rocky outcrops and bluffs lowland to montane areas, NI, SI and Stewart Is., always
terrestrial: *A.a. appendiculatum* inland N, *A.a. maritimum* near coast E.
An upright terrestrial fern with short erect rhizome. Frond to 650 mm; stipe to 200 mm, and
rachis brown underneath, green on top. Lamina elliptic, primary pinnae elliptic, secondary
linear and stubby. Two distinct subspecies *A.a. appendiculatum* has finer pinnae and
lamina. *A.a. maritimum* or Coastal Spleenwort has pinnae rather leathery and stiff. Tertiary
pinnae only on mature fronds, low down. Sori 2–7 mm long, almost marginal.

Polystichum silvaticum Prickly Shield Fern

Cystopteris tasmanica Royal Fern *Blechnum fraseri*

Common Maidenhair

Fan-like Filmy Fern Ground Spleenwort

TRUE FERNS – TWICE OR THREE TIMES PINNATE **241**

Hen & Chicken Fern *Asplenium bulbiferum* **E**
MANAMANA, MAUKU, MOUKI, MOUKU
Lowland to lower montane forests NI, SI, Stewart Is.
A large terrestrial fern with long drooping fronds and stout, erect rhizome. Frond to 1.6 m; stipe, to 400 mm and rachis brown underneath and green on top, flattened towards tips. Lamina elliptic with bulbils, or plantlets, growing from upper surface, these fall off and grow into new plants. Tertiary pinnae indented at tip. Subspecies: *A.b. bulbiferum* as above, *A.b. gracillum* wider lamina, narrower tertiary pinnae and rarely produces bulbils.

Hooker's Spleenwort *Asplenium hookerianum* **E**
ROCKLAX
Clay banks, rocky outcrops, scrub, disturbed forest, lowland and montane areas NI, SI, Stewart Is.
A delicate, upright, terrestrial fern with a short, erect rhizome. Frond to 450 mm; stipe to 200 mm and rachis brown underneath, green on top, scaly. Lamina ovate to rhombic. Primary pinnae quite long, to 80 mm, secondary pinnae stalked. Tertiary pinnae rounded at tips. Two subspecies: *A.h. hookeranium* has rather broader, more rhombic frond while *A.h. colensoi* has narrower fronds and much finer ultimate segments.

Alpine Shield Fern *Polystichum cystostegia* **E**
Upland and alpine areas from Taranaki southwards, including Stewart Is.
A distinctive, quite short, often almost shrubby, light green alpine fern with a short, erect rhizome. Frond to 370 mm; stipe to 120 mm and rachis covered in light yellow-brown scales, especially near base. Lamina narrowly elliptic, secondary pinnae distinctly ovate and deeply divided. Sori mainly on upper pinnae, pale brown indusia.

Black Shield Fern *Polystichum* spp. **E**
COMMON OR SHORE SHIELD FERN, PIKOPIKO, PIPIKO, TUTOKE
Widespread in coastal to montane areas NI, SI and Stewart Is., apart from west coast of SI.
A very dark green, rather prickly, terrestrial fern with a short, erect rhizome. Frond to 850 mm; stipe to 350 mm, and rachis have dark scales, especially near base. Lamina narrowly elliptic to triangular and slightly leathery, secondary pinnae have dark midrib and are deeply divided with pointed apices. Quite variable and previously named *P. richardii*, now *P. oculatum*, *P. neozelandicum* and *P. wawranum*.

Climbing Shield Fern *Rumohra adiantiformis* **N**
LEATHERY SHIELD FERN
Lowland to montane forest, epiphytic on tree ferns, rocks and banks NI, SI, Stewart Is.
A mid-green fern with very large fronds that hang down and a long climbing rhizome which is densely covered with brown scales. Frond to 850 mm; stipe to 350 mm, brown, scaly, rachis green and slightly flattened. Lamina ovate, slightly leathery, lower secondary pinnae are stalked, margins bluntly toothed. First primary pinnae may be 3x pinnate.

THREE OR FOUR TIMES PINNATE

Crepe Fern *Leptopteris hymenophylloides* **E**
HERUHERU
Lowland and montane forest all three main islands, especially damp gullies.
Medium-sized dark green fern with short 200–400 mm trunk; frond to 1.5 m; stipe to 500 mm. Stipe green and slightly hairy. Lamina elliptic to triangular, 3x pinnate, delicate and slightly translucent, especially younger ones. Sporangia scattered on secondary pinnae, not grouped into sori.

Prince of Wales Feathers *Leptopteris superba* **E**
NGUTUKAKARIKI, NGUTUNGUTU KIWI, HERUHERU, DOUBLE CREPE FERN
Moist montane forest from Bay of Plenty southwards, including Stewart Is.
A large, attractive, dark glossy green fern with short, to 1 m, trunk. Frond to 1 m; stipe to 800 mm, light brown. Lamina hairy around mid-rib, 3x pinnate and tapering at both ends. Ends of ultimate pinnae turn upwards to give fluffy appearance.

Hen & Chicken Fern

Hooker's Spleenwort

Alpine Shield Fern

Black Shield Fern

Crepe Fern

Climbing Shield Fern

Prince of Wales Feathers

TRUE FERNS – TWO, THREE OR FOUR TIMES PINNATE **243**

Water Fern *Histiopteris incisa* **N**
MATATA, FALSE BRACKEN
Common open areas lowland to subalpine; roadsides, clearings, forest margins NI, SI, Stewart Is.
Tall, erect, bracken-like, slightly blue-green with long, creeping rhizome. Frond to 1.9 m; stipe to 900 mm, and rachis scaly at base, green when young, maturing yellow. Lamina ovate, primary pinnae sessile, opposite, well spaced, lowest pinnae on secondaries covers rachis. Tertiary pinnae ovate, margins smooth, curled over. Holds water droplets on fronds.

Smooth Shield Fern *Lastreopsis glabella* **E**
Coastal and lowland forest, especially damp areas and stream sides NI, SI, Stewart Is.
Medium-sized terrestrial fern with large, well spaced, pointed, triangular primary pinnae. Rhizome erect. Frond to 650 mm; stipe to 300 mm and rachis brown with scales at base and brown hairs in groove on upper surface. Lamina ovate to triangular, primary and secondary pinnae stalked. Sori on tertiary pinnae.

Diplazium australe **N**
Lowland forest NI and SI, prefers wetter inland areas in SI.
Large, upright, tufted fern with distinctive herring-bone appearance and short erect rhizome. Frond to 1.2 m; stipe to 800 mm. Lamina elliptic to ovate, pinnae well spaced, tertiary pinnae deeply toothed. Sori to 3 mm.

Loxsoma cunninghamii **EG**
Banks, streamsides, clearings in open forest Thames northwards.
Large, quite delicate, pale green fern with erect rhizome. Frond to 600 mm; stipes to 600 mm, mid-brown; lamina ovate, primary pinnae overlapping, tertiary pinnae somewhat notched. Sori on prominent tubular indusia on margins.

Shaking Brake *Pteris tremula* **N**
TURAWERA, TENDER BRAKE
Drier and more open lowland and montane areas, roadsides NI, SI.
Large, handsome, long-stiped, light green fern with short, erect rhizome. Frond to 1.5 m; stipes to 600 mm, brown and shiny. Lamina ovate to triangular. First pinnae long, pointed and triangular, outer segments long and narrow. Sori round margins except tips.

Hymenophyllum bivalve **N**
Lowland to montane forest from Firth of Thames southwards, including Stewart Is.
Distinctive delicate fern, long, thin, creeping rhizome. Frond to 350 mm; stipe to 150 mm, slim and brown, rachis brown, winged. Lamina ovate to triangular, translucent and appearing more branched than pinnate. Pinnae flattened with round apices, slightly toothed. Sori in indusia at tips of ultimate segments.

Irirangi *Hymenophyllum demissum* **E**
DROOPING FILMY FERN, PIRIPIRI
Widespread in moist lowland and montane forest NI, SI, Stewart Is.
Delicate terrestrial or rupestrial and opccasionally epiphytic filmy fern with long, slender, creeping rhizome. Frond to 420 mm; stipe to 170 mm, heavy, rachis narrowly winged. Lamina elliptic or ovate translucent, pinnae clearly veined. Sori in indusia at tips of ultimate segments.

Matua-mauku *Hymenophyllum dilatatum* **E**
Common in lowland and montane forest NI, SI, Stewart Is.; avoids drier areas.
Largest of the filmy ferns, normally epiphytic, or on banks with long, creeping, wiry rhizome. Frond to 450 mm; stipe to 150 mm, heavy, winged and dark brown. Rachis broadly winged and brown. Lamina ovate to narrowly ovate, bright green and translucent. Pinnae rounded with smooth margins. Sori within tips of pinnae covered by indusia.

Smooth Shield Fern

Water Fern

Loxsoma cunninghamii

Shaking Brake

Diplazium australe

Matua mauka

Hymenophyllum bivalve

Irirangi

TRUE FERNS – THREE OR FOUR TIMES PINNATE   **245**

Bracken *Pteridium esculentum* **N**

ARAUHE, RARAUHE, AUSTRAL BRACKEN

Common lowland to subalpine, open areas, pasture, forest margins, roadsides NI, SI, Stewart Is.

Tall, distinctive fern with long, creeping, underground rhizome. Frond to 3 m; stipe to 1 m, and rachis rich brown becoming woody. Lamina broadly ovate, dark green above, lighter below. Primary pinnae long, broad, well spaced. Secondaries similar but narrower, tertiaries very narrow, smooth-margined with edges turned under. Sori marginal around ultimate segments. Rhizome previously an important food source for Maori.

Indescanden Spleenwort *Asplenium richardii* **E**

MATUA-KAPONGA

On rocks, crevices in beech forest and open areas, montane to subalpine habitats NI and SI.

A finely cut, delicate, rich green, terrestrial fern with short, erect rhizome. Frond to 500 mm; stipe to 200 mm, and rachis scaly, brown underneath, green above. Lamina ovate to narrowly ovate. Primary pinnae triangular, secondaries stalked, ultimate segments linear and smooth-margined. Pinnae overlapping. Sori thin, 2–4 mm long.

Lastreopsis microsora **E**

Riverbanks, coastal and lowland forest NI and SI, absent from far south.

A medium-sized, pale green, soft fern with creeping rhizome. Frond to 850 mm; stipe to 350 mm and rachis brown, thin, with flat brown scales and whitish hairs. Lamina ovate to triangular. Primary and secondary pinnae stalked, hairy, ultimate segments rectangular and toothed. Sori on midrib of ultimate segment.

Velvet Fern *Lastreopsis velutina* **E**

Prefers drier areas in coastal and lowland forest NI and SI, in SI only north and east coast south to Dunedin.

Medium-sized, very soft fern with erect rhizome. Frond to 950 mm; stipe to 400 mm and rachis brown with scales at base and hairy further up. Lamina ovate to triangular, brownish green with soft brown hairs. Primary pinnae large and triangular, secondary stalked, ultimate segment rectangular and rounded. Sori on midrib of ultimate segment.

FERN ALLIES

There are three groups of Fern Allies in New Zealand. The *Psilotopsida*, which include psilotum and fork ferns and which lack true roots. The *Lycopsida*, which include clubmosses, spikemosses and quillworts, and the *Equisetopsida* or horsetails. They are all generally more primitive than true ferns with small leaves and sporangia on top rather than on the underside of the fronds.

Psilotum *Psilotum nudum* **N**

Rock crevices from Great Barrier Island to Taupo, thermal regions and Urewera Ranges.

An unusual and distinctive multi-branched almost broom-like plant (frond to 500 mm). Leaves twig-like, dark to bright green with yellow globular sporangia. Prefers rock crevices and can appear like a small shrub.

Long-leaved Fork Fern *Tmesipteris elongata* **N**

Widespread in lowland and montane forest, especially on tree ferns, NI, SI.

An epiphyte with pendulous fronds to 800 mm. Pinnae are arranged spirally and are up to 40 mm in length and generally have a small spine-like tip. Sporangia in pairs at base of upper surface of leaves.

Fork Fern *Tmesipteris tannensis* **E**

Lowland and montane forest, but avoiding drier eastern areas, NI, SI.

Usually epiphytic, pendulous but occasionally erect (stems to 800 mm) and similar to *T. elongata*, but with shorter, spirally arranged leaves (to 30 mm) with long spine-like tips. Sporangia are pointed rather than round-ended.

Indescanden Spleenwort

Bracken

Lastreopsis microsora

Psilotum

Velvet Fern

Long-leaved Fork Fern

Fork Fern

CLUBMOSSES: MATUKUTUKA, WHAREATUA

Lycopodiella cernua N
Lowland and montane forests, Northland to northwestern SI.
An attractive terrestrial plant (to 1 m) with both horizontal and aerial stems, the latter like miniature Rimu trees with drooping ends to the 'branches', which end in strobili or spore cones. Horizontal stems spread over large area, rooting at intervals. Younger branches green, becoming yellow or brown with age.

Alpine Clubmoss *Lycopodium fastigiatum* N
MOUNTAIN CLUBMOSS
Alpine and subalpine areas from Coromandel southwards, including Stewart Is.
Common terrestrial clubmoss of alpine regions (height to 400 mm), bushy at times, distinctive upright, elongated cones to 70 mm. Sterile leaves arranged spirally around stems. Colour varies from green through yellow to chestnut brown.

Creeping Clubmoss *Lycopodim scariosum* N
Montane scrub and subalpine areas, favours damper areas, NI, SI and Stewart Is.
Widespread terrestrial species, height to 500 mm, with alternate leaves lying flat along the stem. Cones are erect, up to 50 mm, unbranched and with few leaves on stem, becoming brown as they mature. Can develop into large patches growing in and over other plants.

Drooping Clubmoss *Huperzia varia* N
IWITUNA, HANGING CLUBMOSS, TASSEL FERN
Widespread lowland, montane and subalpine scrub and forest margins NI, SI and Stewart Is.
Epiphytic and terrestrial forms, former has distinctive drooping stems to 2 m. Terrestrial form upright to 400 mm with characteristic down-curled ends to the branches.

Scrambling Clubmoss *Lycopodium volubile* N
WAEWAEKOUKOU, CLIMBING CLUBMOSS, OWL'S FOOT
Lowland and montane scrub and forest margins NI, SI and Stewart Is.
Widespread terrestrial species with long rambling many branched stems (to 5 m) which are laterally flattened as in *L. scariosum*. Found on open ground or in scrub. Strobili, to 80 mm, hang down in clusters on branched stalks, turning yellow or golden as they mature.

Puakarimu *Lycopodium deuterodensum* N
Lowland and montane forest, regenerating bush and forest margins from North Cape to Lake Taupo.
An erect rather wiry clubmoss with the look of a miniature Rimu. Leaves spreading on some stems, appressed on others. Cones are erect, stalkless, orange-brown and up to 3 cm long. Often in extensive colonies.

African Clubmoss *Selaginella kraussiana* I
Widespread throughout NI and wetter areas of SI.
Bright green with spreading, irregularly branching fronds. Two types of sterile leaves, two lateral rows of larger, 3–5 mm long, leaves and an upper row much smaller, 2–3 mm long, and appressed to the stem, giving it a flattened appearance. Cones on short, to 10 mm, stems. Can form quite large exclusive and invasive mats. Actually a 'Spikemoss', a close relative of clubmosses.

Lycopodiella cernua | Alpine Clubmoss

Creeping Clubmoss | Puakarimu | Drooping Clubmoss

Scrambling Clubmoss | African Clubmoss

FUNGI

Fungi are not plants, but are in a kingdom all of their own, neither plant nor animal. While they are more closely related to animals, mycology, the study of fungi, generally falls under a branch of botany. Fungi come in many shapes and sizes. Most of them are minute, live in the soil and are invisible to the naked eye. There are estimated to be about 1.5 million species of fungus, only about 70,000 of which have been described. In New Zealand there are at least 22,000 species. The fungi that we see are just the reproductive parts of the fungus, the main body, called the mycelium, is a network of fine root-like filaments called hyphae, and is found in the soil or decaying leaves or wood. A mycelium can be very large, covering hectares.

Most of the fungi that you see in New Zealand are Saprobes, these include **Mushrooms**, which have umbrella-shaped fruiting bodies; Polypores or **Bracket Fungi** that stick out from a tree like a shelf; and **Stinkhorns**, which are generally phallic and use rather unpleasant smells to attract flies and help with their propagation. Some Saprobes develop spores on their smooth outer surfaces, these include **Crust Fungi**, **Coral Fungi** and **Cup Fungi**, all fairly self-descriptive. **Sooty Moulds** are also saprobic fungi, these are especially common on beech trees, where they are often found with the Sooty Beech Scale *Ultracoeloscoma* spp., which produces honeydew.

Most fungi that you see will be native and many endemic, however those found in the open, especially near alien tree species, are more likely to be introduced ones. Beech forests are particularly good places to search for fungi. Specific identification of fungi is not easy without a detailed guide and careful examination. As always, while some species are edible, do not under any circumstances eat any unless you are sure of their identity.

MUSHROOMS OR GILL FUNGI

These umbrella-shaped fungi have gills on the underside of the head or cap. One of the largest and most easily identifiable is an introduced species, the **Scarlet Flycap** *Amanita muscaria,* with its white-dotted, red top. These are particularly associated with introduced conifers. Much smaller are several species of **Mosscap** *Rickinella* spp.; these can be quite brightly coloured and generally found in grasslands. There are a wide variety of gill fungi, growing mainly on rotting wood. The introduced **Orange Poreconch** *Favolaschia calocera* is widespread and often grows in profusion. Parasitic species include several species of *Amarillia* which can be quite large or delicate and translucent, they are also often brightly coloured, red, yellow, orange or blue and may grow individually or in great clumps. Another parasitic species is the **Tree Swordbelt** *Agrocybe parasitica* which when mature has a distinctive collar under the cap.

CUP FUNGI

These distinctive fungi have quite small cup-shaped fruiting bodies. The **White Birdsnest** *Nidula candida* is a good example of this type. They are often brightly coloured and have the spores on the upper surface rather than underneath the cap.

CLAVARIA

These fungi, found on rotting wood, are easily recognisable but hard to identify specifically, they look very much like bean or pea sprouts, though some are quite brightly coloured. *Clavaria* spp. are generally simple, *Clavicorona* spp. are branched with crown-shaped branch ends, and *Calvulina* spp. may be either simple or branched.

POUCH FUNGI & PUFFBALLS

The main feature of these fungi is that they store the spores within a globular fruiting vessel, this then splits open to release the spores. **Puffballs** *Lycoperda* spp. are largely spherical and release their spores either through a single central vent, or by simply bursting open. **Earthstars** have an outer skin that peels back to reveal the spore receptacle inside. In contrast, **Pouch Fungi** can be very oddly and irregularly shaped and split open irregularly too. They can be confused with Puffballs, but when cut open they are sponge-like and moist inside whilst Puffballs are dry. A good example is *Gyromitra tasmanica* which looks a bit like a deflated football. Some pouch fungi are like truffles and are found underground.

Scarlet Flycap   *Gyromitra tasmanica*

Amarillia sp.   Tree Swordbelt

Orange Poreconch   White Birdsnest

STINKHORNS

These distinctive fungi can be quite long, up to 200 mm, and narrow; many have a distinctly phallic appearance. The tissue surrounding the spores at the tip of the fungus turns into a gelatinous slimy substance which emits a foul smell that attracts flies and other insects, which then help to disperse the spores. A variation on this is the **Tentacled Stinkhorn** *Aseroe rubra* which produces long scarlet tentacles to produce and distribute its spores.

JELLY FUNGI

These are, as their name suggests, soft and floppy. They come in various shapes and colours, generally on wood. A common one is **Wood Ears** or **Hakeke** *Auriculana cornea*, which is found on dead trees. They have no stem and vary in colour from pale brown through to mid or even dark grey. They have white spores.

CRUST AND BRACKET FUNGI

Crust Fungi can look rather like a large lichen, varying in colour from yellow to red. **Bracket Fungi** look like small shelves stuck on a tree trunk. They are often very large, sticking out up to 300 mm from a tree and being 100 mm or more thick. **Artist's Porebracket** *Ganoderma applanatum* is a good example of this. They are generally dark on top and white, where the spores are, underneath. Another which is widespread, especially on dead trees, is *Coriolus versicolor* with often quite dramatic varicoloured growth rings. The **Orange Porebracket** *Pycnoporus coccineus* is bright orange.

BRYOPHYTES – MOSSES, LIVERWORTS AND HORNWORTS

Bryophytes are plants that lack vascular tissue that circulates liquids. Like ferns, they do not have flowers or seeds but reproduce via spores. Bryophytes are divided into three groups: Mosses (*Bryophyta*), Liverworts (*Marchantiophyta*) and Hornworts (*Anthocerotophyta*). New Zealand has well over 1000 native species, many are endemic and there are relatively few introduced species apart from in urban areas. The sporangia that hold the spores are often on long slender stalks.

MOSSES

New Zealand mosses can be divided into three 'types', **true mosses**, which includes the bulk of the 550 species found here; **sphagnum** or **peat mosses**, nine species, and **granite mosses**, found in the alpine region, five species. Around 20% are endemic. Mosses are amazingly resilient plants and occupy a very wide range of habitat, from the coast to alpine areas. They generally prefer cool, damp, shady locations and can be found growing on the ground, on rocks or as epiphytes in trees. They are often quite small and prostrate, though they can grow laterally into extensive patches. While most mosses are green, some, especially the sphagnums, are yellowish or even pink.

Mosses reproduce sexually and have a two-stage cycle. The male sex organ, called antheridia, produce antherozoids, or sperm, which must travel through a film of water to the female organ to fertilize the egg. The fertilised egg produces a sporophyte or spore capsule, often on a long slender stalk, this releases spores which germinate to form new plants. Some species have male and female sex organs, others have them on separate plants. In some species, the male plants are minute and grow on the female plant.

Moss identification is rarely easy and examination with a good hand lens or even a microscope is often needed to get beyond the genus level. **Umbrella mosses**, *Hypnodendron* spp. and *Hypopterygium* spp., are a good example with two genera, and several species of each genus. It is quite easy to recognise an umbrella moss, but species identification is much harder. A few mosses are quite easy to identify; *Dawsonia superba*, the largest moss in New Zealand, can grow to 500 mm and at first glance can be mistaken for a small pine seedling. Another upright moss of the forest floor which has a real tongue-twister of a name is *Dendroligotrichum dendroides*. It is smaller than *Dawsonia* and has branched fronds and is a richer green. If you see long stringy mosses festooning the trees, this is quite likely **Old Man's Beard** *Weymouthia mollis*, which forms attractive pale green 'curtains' hanging down from the branches of trees. *W. cochlearifolia* which has the same

Tentacled Stinkhorn

Artist's Porebracket

Wood Ears

Orange Porebracket

Jelly Fungus

Coriolus versicolor

Dawsonia superba

Dendroligotrichum dendroides

Old Man's Beard

common name, forms a rather neater and better trimmed beard. Both species prefer moister areas, which generally means west coast or montane forest.

In upland areas and around ponds, sphagnum mosses *Sphagnum* spp., the classic bog moss, are common but are also hard to identify to species level. They form extensive, thick, very soft cushions which over time decay into peat.

If you encounter a large, round cushion-like moss it is quite likely a species of *Bryum*, and especially in wetter forest areas you will come across the delightfully named **Pipe-cleaner Moss** *Ptichiomnion aciculare*, with its somewhat erect individual fronds.

Other common mosses are the attractive soft mosses *Dicranaloma*, often found mixed with other moss and liverwort species and the **Milk Moss** *Leucobryum candida*, with its distinctive whitish appearance. *Hypnodendron arcuatum* is an attractive green moss with branched fronds which is often found clothing a steep bank or cliff, a site also favoured by *Cyathophorum bulbosum*, which has longer but unbranched fronds.

LIVERWORTS

The name means 'liver herb' and derives from the old belief that because many species had rather liver-like shapes, they would be good for treating liver ailments, a belief with no foundation. Liverworts are very old. They have been around for some 400 million years, far longer than ferns, mosses and flowering plants. With over 500 species, New Zealand is home to around 7% of all species of liverwort, a testament to the cool, damp, dark forests that originally covered much of the country and provided them with an ideal habitat. They are immensely varied in form and size, from quite large sheet-like ones such as *Monoclea forsteri*, which can be up to 200 mm across, to ones that are almost invisible with fronds only 0.3 mm thick.

Liverworts are closely related to mosses and are often hard to distinguish from them, however they do not have a vein or nerve along the middle of the leaf, while many mosses do. They have two or three rows of leaves while mosses have them in a spiral, though this may be flattened. Liverwort leaves are more diverse and leaves with lobes, especially three or four lobes, are liverworts, mosses almost always have simple leaves. Liverworts are more colourful, so red, magenta, gold, etc., are more likely to be liverworts.

Liverworts come in two 'types': leafy ones, which have leaves and are easily confused with mosses, and thalloid ones which are made up of sheets of cells. Three common genera which are hard to distinguish are *Schistochila*, *Lepidopzia* and *Solenostoma*, so identifying to species level is extremely difficult. *Schistochila appendiculata* is a common species which is often found on shady, damp banks, it is bright green and has a rather flattened frond and frequently a very glossy, moist look. Another distinctive species is *Marchantia foliacea* with its dark green, glossy, lobed fronds or leaves.

Liverworts reproduce both sexually and asexually or vegetatively. Thalloid liverworts of the genus *Marchantia* have small 'splash cups' on their surface containing tiny green 'gemmae', not quite seeds, more like proto-plants. These gemmae are knocked out of the cups by raindrops and may end up a metre or more away, where they grow into new plants.

Liverwort or Moss? Moss leaves are generally arranged radially around the stem, while liverworts are more pinnate with leaves on either side. Most mosses have a central vein or midrib, most liverworts do not. Lobed leaves are a liverwort indicator, as are spore capsules that split open, moss capsules are end-opening.

HORNWORTS

Very similar to Thalloid liverworts, but with distinctive tapering sporophytes 20–50 mm long with spores developing along their whole length. There are only nine species in New Zealand, of which the genus *Anthoceros* is the commonest.

Sphagnum bog   *Sphagnum* sp.

Milk Moss

Pipe-cleaner Moss

Cyathophorum bulbosum

Bryum sp.

Dicranaloma sp.

Schistochila appendiculata   *Monoclea forsteri*

LICHENS

Lichens (pronounced *'likens'*) are tough, adaptable and very widespread; they occur in virtually all terrestrial habitats and are very varied in shape, form and colour. Lichens are very long-lived, colonies of *Rhizocarpon alpicola* in Lapland are more than 4000 years old. They grow very slowly and steadily to the extent that they have given rise to the science of Lichenometry or measuring the age of an exposed surface by the lichens growing on it. This is useful for lava flows, glacial activity, sea level changes etc.

Lichens are not a single 'plant' but a symbiotic partnership between a fungus, which is itself not a plant, and a green or photosynthesizing alga or a cyanobacterium, occasionally both. The fungus provides the alga or cyanobacterium with a 'home' and with minerals and water which the alga uses to provide food, via photosynthesis, for itself and the fungus. The partnership is very stable and almost certainly extends the range of both partners. The visible structure is the fungus, the algal partner being within this structure. The fungal partner of a lichen is known as a 'lichenising fungi' and can only survive in partnership with its algal partner, whereas the algal partner can survive independently. There is a school of thought that thinks the fungi are at least partly parasitic on the algae.

New Zealand is rich in lichens, with probably over 2000 species, however only around 40% are endemic. They are found in all environments from the coast to the snow line, even around volcanic vents, but are susceptible to industrial pollution. Identification frequently requires a microscope or even chemical reagents. Given that there are so many species, this section is just to help you identify the three basic types and to introduce you to the wonders of the lichen world.

CRUSTOSE OR CRUSTING LICHENS

These, as their name suggests, form a covering or crust over the surface on which they are growing, often a rock or tree, others prefer walls, roofs or even roads. Some are brightly coloured, red or yellow, others are bicoloured such as the dramatic black and yellow *Rhizocarpon geographicum*. Many are almost circular and grow outwards at a very steady measurable rate, at the same time recording variations in atmospheric conditions.

FOLIOSE OR LEAFY LICHENS

This group of lichen produces quite large leaf-like fronds, often 300 to 400 mm long, and frequently lobed. One of the commonest, *Pseudocyphellaria homoeophylla*, is bright green and can be found in profusion on tree trunks in wetter areas. Other species are grey, or brown, or as with *Hypogymnia enteromorpha*, black and white.

FRUTICOSE OR SHRUBBY LICHENS

These lichens are again quite distinctive, looking like miniature bushes or shrubs. They may be either erect on the ground or in trees or hanging, often in trees. Their structure can be quite intricate as with the coral or lace lichens of the genus *Cladina* and *Cladonia* and various species of *Usnia*, sometimes referred to as old man's beard

REPRODUCTION

Most lichenising fungi are **Ascomycetes** or cup fungi thanks to their clearly visible reproductive cups (these produce spores which are washed out by raindrops). The spores then germinate but need to find an algal partner quickly in order to survive. Most lichens can also reproduce vegetatively by dropping off small portions of the thallus containing the algal partner. This is particularly the case with fruticose lichens where parts of the thallus break off when dry.

Crustose Lichens

Crustose Lichens

Rhizocarpon geographicum

Pseudocyphellaria homoeophylla

Hypogymnia enteromorpha

Cladonia sp.

Usnia sp.

PLACES TO VISIT

This section is a brief guide to places to visit to observe all the species, and more, that are mentioned in this book. By no means exhaustive, it does provide a good start. I would recommend visiting the local information sites that can be found in most towns.

NORTH ISLAND

Coast and Wetland

There are a number of large estuaries and harbours that are excellent for birdwatching, especially in the summer when migrant waders are present in large numbers (high tide is always the best time to observe them). **Miranda** on the Firth of Thames is probably the best of all, and includes one of only three Chenier beaches in the Southern Hemisphere, but smaller estuaries and harbours like **Maketu** and **Ohiwa** in the Bay of Plenty can be surprisingly rich and sometimes more accessible. Two gannet colonies at **Muriwai** near Auckland and **Cape Kidnappers** in Hawke's Bay are amazing during the breeding season from August through March. The fenced peninsula at **Tawharanui** near Warkworth is also worth a visit. There are three island sanctuaries you can visit quite easily that are all rich in bird life: **Tiritiri Matangi** near Auckland and **Kapiti** near Wellington both require booking in advance, but there is no need to book to visit **Somes Island** in Wellington Harbour. It is also possible to visit and stay on **Mayor/Tuhua Island** in the Bay of Plenty. Wetlands worth a visit are **Whangamarino** near Meremere and **Kaituna** next to Maketu Harbour in the Bay of Plenty.

Lowland and Montane Forest

Trounson Kauri Park in Northland is the best place to see the awesome Kauri trees, otherwise **Te Urewera National Park** and **Whirinaki**, **Pureora** and **Pirongia Forest Parks** all offer wonderful forest walks, mainly though broadleaf–podocarp forest, where you can observe the remarkable diversity of the bush. Most of the other protected forest areas also offer good bush walks. **Otari Wilton Bush** in Wellington is an excellent and easily accessible area of bush close to the city centre, as is the excellent **Karori Wildlife Sanctuary**. The **Waitomo Caves** near Otorohanga are wonderful for glow worms.

Alpine and Subalpine Regions

Tongariro and **Egmont National Parks** are the only two areas with genuine alpine habitat in the north. Elsewhere there is subalpine vegetation on the tops of many of the ranges.

Lakes and Rivers

Lake Taupo and the many lakes in and around **Rotorua** offer a number of good opportunities for birdwatching, probably rather better in the winter. The smaller lakes with fewer boats are generally better.

Offshore Boat Trips

These can offer good opportunities to see seabirds and cetaceans. Trips can be arranged in a number of places including Whangarei, Auckland, Tauranga, Whakatane and Napier.

Volcanic and Geothermal Features

The central North Island's **Volcanic Plateau** and the Rotorua–Taupo region has a wide variety of volcanic features and thermal attractions. I particularly like Waimangu, a valley created one night in 1886 by the explosion of Mount Tarawera.

For a spectacular day out try **White Island**, off the coast near Whakatane, which is the most active volcano in New Zealand and offers wildlife opportunities on the boat trip there and back. Mayor/Tuhua Island, also in the Bay of Plenty, offers the best obsidian flows in New Zealand.

SOUTH ISLAND
The South Island has more wilderness area, but much of it is not very accessible unless you are into serious tramping.

Coastal and Wetland
There are fewer large estuaries and harbours than in the north, but the **Wairau Estuary** and **Big Lagoon** near Blenheim are good, and there are opportunities around **Lake Ellesmere** near Christchurch, **Okarito Lagoon** on the West Coast and on the coast between Bluff and Fortrose in Southland. **Taiaroa Head** near Dunedin has the only mainland breeding colony of the Northern Royal Albatross. For cetaceans and seabirds, there is no better place than **Kaikoura** on the east coast between Blenheim and Christchurch. **Farewell Spit** in Golden Bay has a large Gannet colony and other interesting wildlife. **Oamaru** hosts Little Blue and Yellow-eyed Penguins, and all around the south coast there are colonies of Hooker's Sea Lion and New Zealand Fur Seal.

Lowland and Montane Forest
Much of the forest is Beech, especially further south and higher up. Most is already within the National Park system but the **Catlins Forest Park** in the south and **Richmond Forest Park** in the north also have good examples of native forest.

Lakes and Rivers
The braided rivers of the South Island are an amazing sight. The **MacKenzie Basin**, especially the Ahuriri Valley, is excellent and varied offering braided rivers, wetlands and glacial features. The Black Stilt recovery centre in **Twizel** is your best chance of seeing this rarest of birds.

Alpine
The South Island has very large areas of alpine habitat. Easily accessible ones include the **Cobb Valley** and **Mount Arthur** in the north, **Arthur's Pass** and **Homer Tunnel** further south, plus many others in the various national parks and protected areas. **Mt Cook National Park** is very accessible and offers easy access to alpine flora and fauna as well as glacial features.

Geology
Try to visit the Karst limestone formations on **Takaka Hill**, near Nelson and **Fox** and **Franz Josef glaciers** on the West Coast. Spot the Alpine Fault near the glaciers and in the north. Also try to experience the amazing glacial scenery around **Aoraki Mount Cook**. **Curio Bay** in the Catlins has a wonderful fossilized forest and the amazing **Moeraki Boulders** are on the coast a bit further north .

MAINLAND ISLANDS
New Zealand has pioneered the concept of creating 'inland islands' – areas where all or most pest species have been eradicated to enable the natives to flourish. These fall into two categories, the fenced and the unfenced.

Of the fenced islands probably the best-known inland sanctuary is **Karori Sanctuary**, a 2.52 km² reserve only five minutes from the centre of Wellington. **Tawharanui Open Sanctuary** is a fenced-off peninsula north of Auckland, near Warkworth, which is open to the public. Coastal and wetland habitat. **Maungatautari Ecological Island** has the longest pest-proof fence of all at 47 km. It surrounds an isolated 800 m volcanic cone on the Waikato plain south of Hamilton. **Bushy Park** is an area of lowland native rainforest near Wanganui with a 4.6 km pest-excluding fence. The **Mount Bruce National Wildlife Centre** near Masterton is a breeding centre for endangered species including kaka, kokako, brown kiwi, stitchbird and yellow-crowned parakeet. There are also walks through ancient podocarp forest and easily viewed Tuatara. **Orokonui Eco-sanctuary** is just north of Dunedin in the South Island and is a 3 km^2 sanctuary with a 10 km pest fence around it.

Two other smaller fenced sanctuaries are **Riccarton Bush** in Christchurch, and **Young Nick's Head** in Hawke's Bay. A number of others are being planned. All are very well worthwhile visiting.

Unfenced mainland islands are created by an intensive pest eradication campaign, and their pest-free status is maintained by a long-term extensive monitoring programme. **Ark in the Park** is a large native species recovery project in the Waitakere Ranges to the west of Auckland covering some 20 km^2. **Hinewai** is a private reserve covering 12.9 km^2 on the Banks Peninsula near Christchurch. It runs from close to the coast to a subalpine peak. The **Rotoiti Nature Recovery Project** focuses on an area of honeydew beech forest in the Nelson Lakes National Park.

DOC runs a programme of six Mainland Islands, all unfenced, in the North Island these are: **Trounson Kauri Park** in Northland; the **Northern Te Urewera Ecological Restoration Project**; **Paengaroa**, a small 1.17 km^2 reserve near Taihape, and **Boundary Stream** near Napier. In the south, there is the Rotoiti Lake project mentioned above and, the largest, the **Hurunui Mainland Island** in North Canterbury covering more that 120 km^2.

ISLAND SANCTUARIES

There are a large number of islands around the coast, varying from small islets to Great Barrier Island at 285 km^2 and a height of 621 m. These islands are refuges for wildlife and have since been used as part of an extensive and sophisticated recovery programme. Many of these offshore nature reserves, such as Codfish and Little Barrier islands, are off limits without a special permit, but Tiritiri Matangi, Kapiti, Somes and Mayor/Tuhua, all mentioned above, are accessible to visitors.

MARINE RESERVES

There is a network of more than 30 marine reserves and protected areas around New Zealand. These are an important part of the overall conservation strategy of the country, especially given the popularity of boating and fishing in New Zealand. No fishing is allowed within marine reserves, but they can be an excellent place for snorkelling. Protected marine areas also include the **Nga Motu/Sugar Loaf Islands Marine Protected Area** in Taranaki and three Marine Parks: **Mimiwhangata** in Northland, **Tawharanui** just north of Auckland, and, by far the largest, the **Hauraki Gulf Marine Park** that covers an area of more than 1.2 million hectares on the east coast of the Auckland and Waikato regions. It includes more than 50 islands, several of them wildlife reserves or sanctuaries, and five marine reserves.

NATIONAL PARKS AND PROTECTED AREAS

National Parks cover more than 30,000 km² and are a key element of the overall conservation strategy. New Zealand's 14 National Parks are listed below, in rough order north to south.

When visiting, remember that you are not allowed to remove anything but rubbish from the parks: **Leave only frootprints, take only pictures and memories.**

Te Urewera
This is the largest area of native forest in the North Island. A remote and wild area, it features lakes Waikaremoana and Waikareiti and large areas of hills clad with native forest, much of it beech. There is good birding, especially for Kokako, and over 650 species of flowering plant plus ferns, mosses, lichens and fungi.

Tongariro
Sacred to Maori, the three active volcanic peaks that make New Zealand's first National Park were gifted to the nation in 1887 by Te Heuheu Tukino IV, paramount chief of Ngati Tuwharetoa. The park is an excellent area to observe alpine and subalpine habitats. Geologically, the park is interesting for its many amazing volcanic features and wonderful views.

Whanganui
A lowland park on the Whanganui River. Good for podocarp–broadleaf forest with rata, rimu, tawa and rewarewa prominent, and beech forest on the ridges. The river cuts through soft mudstone, creating spectacular gorges, plus there are some interesting subfossil deposits. A good area for birds and plants.

Egmont
This park is centred on the spectacular snow-capped volcanic cone of Mount Taranaki or Mount Egmont (as it is also known). Walk from lowland and montane podocarp–broadleaf rainforest through subalpine bush to alpine herbfields and fellfields and the snow line. Good birdlife, wonderful alpine plants and some good bogs too.

Abel Tasman
The smallest national park, it is noted for its coastline, especially the beaches, inlets and estuaries. Much of the bush is regenerating. There is both beech and podocarp–broadleaf forest and areas of geological interest, including granite and limestone outcrops, the latter linked to the Karst Limestone formations found inland on Takaka Hill.

Kahurangi
At over 4,500 km², it is one of the largest parks containing a wide variety of habitats and wildlife. Excellent for alpine and subalpine flora and fauna, it is also the **last major refuge of the huge *Powelliphanta* snails and one of their favourite prey items, the native earthworm which can be up to a metre long.**

Paparoa
A small park on the West Coast centred on the remarkable limestone pancake rocks at Punakaiki. Inland, the vegetation is a lush temperate rainforest with good examples of podocarp–broadleaf forest along with Nikau Palms.

Nelson Lakes
Situated at the northern end of the Southern Alps, it is good for montane forest, especially beech and alpine habitat plus the lakes Rotoiti and Rotoroa. The park has a wide variety of bird and animal life.

NATIONAL PARKS, PROTECTED AREAS AND PLACES TO VISIT

Bay of Islands

Mimiwhangata Marine Park

Poor Knights Is.

Trounson Kauri Park • Whangarei •

Whangarei Harbour

Hauraki Gulf Marine Park

Little Barrier I.

Great Barrier I.

Cape Rodney

Tawharanui Marine Park

Kaipara Harbour

Tiritiri Matangi I.
Long Bay

Motu Manawa Pollen
Muriwai

AUCKLAND

Hauraki Gulf Marine Park
Te Whanganui-a-Hei

Coromandel FP

Manukau Harbour

Miranda • Thames

Tuhua (Mayor) I.

Waikato River

Whangamarino
Wetland

Bay of Plenty

Te Paepae o Aotea
• White I.

HAMILTON

Tauranga
Maketu

Pirongia FP

Whakatane

Kawhia Harbour

Maungatautiri

Ohiwa

Raukumara

Waitomo
Caves

Rotorua

Pureora FP

Whirinaki FP

TE UREWERA
NP

Parininihi

Nga Motu/Sugar Loaf Islands
Marine Protected Area

Lake
Taupo

Te Tapuwae
o Rongokak

Gisborne
Young
Nicks
Head

Tapuae •

New Plymouth

Volcanic
Plateau

EGMONT NP

Kaweka FP

WHANGANUI
NP

TONGARIRO NP

Hawke Bay

Napier

Taihape •

Cape Kidnappers

Ruahine FP

Whanganui

Te Angiangi

Tararua FP

Mt Bruce

Kapiti Island

Masterton

WELLINGTON

Taputeranga

Aorangi FP

Somes I.
Rimutaka FP

Fiordland Marine Rese

RA

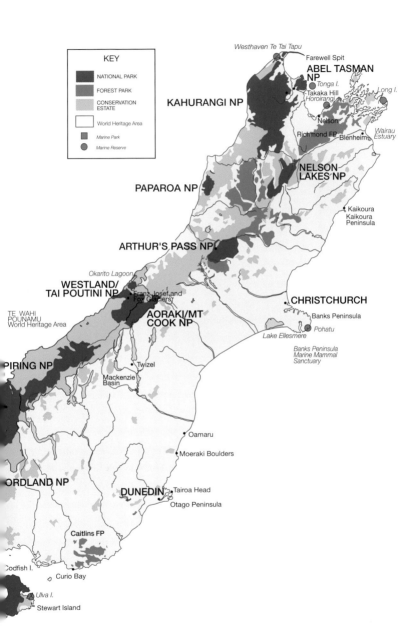

KEY

NATIONAL PARK

FOREST PARK

CONSERVATION ESTATE

World Heritage Area

Marine Park

Marine Reserve

Westhaven Te Tai Tapu

Farewell Spit

ABEL TASMAN NP

Tonga I.

Takaka Hill

Horoirangi

Long I.

KAHURANGI NP

Nelson

Richmond FP

Blenheim

Wairau Estuary

NELSON LAKES NP

PAPAROA NP

Kaikoura

Kaikoura Peninsula

ARTHUR'S PASS NP

Okarito Lagoon

WESTLAND/ TAI POUTINI NP

Franz Josef and Fox Glaciers

CHRISTCHURCH

Banks Peninsula

TE WAHI POUNAMU World Heritage Area

AORAKI/MT COOK NP

Pohatu

Lake Ellesmere

Banks Peninsula Marine Mammal Sanctuary

PIRING NP

Twizel

Mackenzie Basin

Oamaru

Moeraki Boulders

ORDLAND NP

DUNEDIN

Tairoa Head

Otago Peninsula

Caitlins FP

Codfish I.

Curio Bay

Ulva I.

Stewart Island

Arthur's Pass

A mountainous park straddling the Main Divide, with beech forest to the east and podocarp–broadleaf to the west. Braided rivers, alpine plants and wildlife make this one of the best and most accessible parks.

Aoraki Mount Cook

A high alpine and mountainous park including Aoraki/Mount Cook, which at 3,754 m is the highest peak in New Zealand. The park also features the two longest glaciers in New Zealand; the Tasman and Hooker. Little forest, but wonderful alpine uplands with a wide range of flora and fauna.

Westland–Tai Poutini

This park is on the west of the Main Divide and includes the Fox and Franz Joseph glaciers, which are accessible by easy trails. Vegetation includes alpine meadows and fellfields, beech and podocarp–broadleaf rainforest as well as coastal wetlands and lagoons. A huge variety of flora and fauna.

Mount Aspiring

A very mountainous park centred on the 3,030 m Mount Aspiring and great for trampers. The vegetation is beech forest and alpine with a large variety of alpine flora and fauna.

Fiordland

The largest National Park, but with very limited access. Vegetation varies from alpine to coastal rainforest with rainfall up to 12 m a year. There is a wide variety of plants, over 700 species, as well as birds and invertebrates including, annoyingly, the sandfly or namu.

Rakiura

Rakiura covers 85% of Stewart Island. It is largely podocarp–broadleaf forest, with subalpine vegetation on the highest areas. There are extensive wetlands in the middle of the island and large coastal dune systems to the west. It is ecologically quite distinct.

PROTECTED AREAS

There are 14 Forest Parks in the North Island, all in upland areas, apart from Northland Forest Park which includes the finest remaining Kauri trees. Coromandel, Kaimai–Mamaku, Whirinaki and the Raukumara around the Bay of Plenty offer a wide variety of native forest. Pirongia and Pureora to the west contain large areas of podocarp–broadleaf forest and an excellent and varied wildlife.

South Island has seven Forest Parks and four Conservation Parks. Among these are Craigieburn and Lake Sumner, which offer wonderful upland and alpine experiences, and the Catlins on the south-east coast which has some of the most unspoilt rainforest in the country. In addition, it is right next to a wild and fascinating coastal area.

There are a large number of other conservation areas on both the North and South Islands — Conservation Parks, Scenic Reserves and so on, all of which provide an element of wilderness and offer wildlife-watching opportunities.

World Heritage Sites

Te Wahipounamu World Heritage Site encompasses the Aoraki Mt Cook, Westland–Tai Poutini, Mt Aspiring and Fiordland national parks. **Tongariro National Park** is a World Heritage Site listed for both its cultural and natural heritage values.

GLOSSARY

Adnate   Attached by the whole width, lacking a stalk.

Alien   Non-native species introduced into the ecosystem.

Apices   Plural of apex.

Anther   Pollen bearing part of a stamen.

Axil   Angle at the junction of a leaf stalk and stem.

Bifoliolate   Having two leaves.

Bract   A modified, usually reduced, leaf often associated with flower stems.

Bulbil   Small plantlet that grows on a frond, drops off and becomes self-supporting.

Calyx   Refers to sepals as a whole, usually when joined covering a bud.

Caruncle   A fleshy naked outgrowth – as in chicken crest.

Cephalothorax.   Fused head and thorax in spiders.

Crenulated   Having a margin of small rounded teeth.

Diurnal   During daytime – as opposed to nocturnal.

Divaricating plant   Branches spread apart at a wide angle. Refers to a group of NZ shrubs and low-growing trees with small leaves and interlaced, wiry stems.

Drupe   Fleshy fruit containing one or two seeds.

Endemic   Found naturally only in that location, or breeds only in that location.

Epiphyte-epiphytic   Growing on another plant or structure, e.g., rock or building.

Fellfield   Alpine, rather rocky habitat.

Frond   Complete leaf of fern including, stipe, lamina.

Gamete   Sex cell which joins with another of opposite sex in sexual reproduction.

Gametophyte   Plant that produces the gametes.

Herbfield   Alpine grasslands often with numerous perennial herbaceous plants.

Indusium/a   Thin outgrowth of tissue covering the sorus.

Inflorescence   Any arrangement of more than one flower.

Instar   A stage of an insect or other arthropod between moults.

Invasive   Introduced or alien species that become established, changing the ecosystem.

Lahar   Mudflow composed of volcanic material and water.

Lamina   Flattened blade or leafy part of a fern frond.

Lanceolate   Lance-shaped, tapering from a rounded base towards an apex.

Larva   Immature or juvenile stage of insect, e.g., caterpillar.

Lek   System of courtship where males display, often competitively, to attract females.

Lores   Area immediately in front of the eye on a bird.

Montane   Hilly or mountainous.

Morph   Form, plumage variation.

Mustelid   Family of mammalian carnivores including stoats, weasels, ferrets.

MYA   Million years ago.

Native   Found naturally in that location, not introduced.

Nocturnal

Obovate/Oboviod   Egg-shaped with narrow end at base.

Ovate/Ovoid   Egg-shaped, with a tapering point.

Opercula   Bony plate in molluscs that fills opening when animal withdraws.

Palmate   Hand shaped, fingered.

Panicle   A branched raceme with each branch bearing a raceme of flowers.

Parthenogenesis   Development of an egg without fertilization.

Pensile   Hanging freely.

Petiole   Leaf stalk.

Phylloclade   Flattened stem or branch that looks like, and serves the function of, a leaf.

Pinna/ae   Segment of a divided lamina or leaf (pinnae plural of pinna).

Pinnate   Having a lamina or leaf divided into pinnae.

Pistil   Female organ of a flower, includes stigma, style and ovary.

Pneumataphores   Plant roots that stick up above the surface and can 'breathe'.

Podocarp   Very tall coniferous tree of Podocarpaceae family.

Polyanandrous   Mating of female with more than one male in any one breeding season.

Raceme   Inflorescence with central 'stem' with flowers on unbranched stalks.

Rachis Main stem of fern from lowermost pinna to apex.

Recurved Curved backwards or downwards.

Revolute Rolled downwards from tip or margins to the undersurface.

Rhizome Underground stem, may also appear above ground.

Rictal Hairs around Kiwi bill.

Rupestral Growing on rock.

Sessile Stalkless, attached directly at the base.

Sorus/Sori A cluster of two or more sporangia.

Spathulate Flattened, spatula or sword shaped.

Speculum Well defined patch of feathers on secondaries or secondary coverts.

Sporangium/a Capsule that contains spores.

Spore Single celled reproductive unit, similar to seed in flowering plants.

Sporophyte Spore-producing plant in ferns and fern allies.

Stamen Male part of flower, produces pollen.

Stigma Female part of flower that receives pollen.

Stipe Stalk of a frond.

Stipule Scale-like appendage at base of the stipe, often in pairs.

Stolon/Stoloniferous Slender lateral stem growing out of main rhizome.

Stroboli Cone-like structure.

Sulcus Upper edge of lower mandible on a bird bill or beak.

Stool Pedestal or trunk of a fern.

Subnival Below the snow-line.

Thallus Simple plant body, not split into stems, leaves, etc.

Thorax Mid section of insect body, between head and abdomen.

Tomentose Covered with short, dense, matted hairs.

Umbel Umbrella-shaped cluster of flowers.

Umbellifer Plants with umbels.

Ungulate Hoofed grazing mammal, e.g., goat, deer, sheep.

Unifoliate With a single leaf.

Viviparous Giving birth to live young.

USEFUL WEBSITES AND CONTACTS

Ark in the Park
www.forestandbird.org.nz/what-we-do/projects/ark-in-park

Auckland Regional Parks
www.arc.govt.nz/parks/

Bushy Park
www.bushypark.org.nz
info@buskyparksanctuary.org.nz /
Tel: 06 342 9879

Conservation Volunteers New Zealand
www.conservationvolunteers.co.nz
info@conservationvolunteers.co.nz

Department of Conservation
www.doc.govt.nz
National Office (Wellington)
enquiries@doc.govt.nz / Tel: 04 471 0726

ECO – Environment and Conservation Organisations of NZ
www.eco.org.nz / eco@reddfish.co.nz
Tel/Fax: 04 385 7575

Greenpages
www.greenpages.org.nz
Greenpages lists a large number of local and national conservation organisations.

Greenpeace Aotearoa New Zealand
www.greenpeace.org.nz
info@ greenpeace.org.nz
Tel: 04 630 6317

Hinewai Reserve – Maurice White Native Forest Trust
Long Bay Rd, RD3, Akaroa 7853, New Zealand
Tel: 03 304 8501

Karori Sanctuary Trust
www.sanctuary.org.nz
kwst@sanctuary.org.nz / Tel: 04 920 9200

Maungatautari Ecological Island Trust
www.maungatrust.org
mail@maungatrust.org / Tel: 07 823 7455

Miranda Shore Bird Centre
www.miranda-shorebird.org.nz
info@miranda-shorebird.org.nz
Tel: 09 232 2781

New Zealand Native Forests
Restoration Trust
www.nznfrt.org.nz

New Zealand Ecological Restoration
Network Inc.
www.nzern.org.nz / Tel: 03 980 0902

Orokonui Ecosanctuary – Otago Natural
History Trust
www.orokonui.org.nz
info@orokonui.org.nz / Tel: 03 482 1755

Ornithological Society of New Zealand
www.osnz.org.nz
OSNZEO@slingshot.co.nz

Pukaha Mount Bruce
National Wildlife Centre for endangered
species breeding
www.mtbruce.org.nz
info@pukaha.org.nz / Tel: 06 375 8004

QEII National Trust
www.openspace.org.nz
info@openspace.org.nz / Tel: 04 472 6626

Royal Forest and Bird Protection Society
www.forestandbird.org.nz
office@forestandbird.org.nz
Tel: 04 385 7374

Seaweed Association of New Zealand
(SANZ)
www.sanz.org.nz
seaweed@wave.co.nz / Tel: 07 862 8424

Supporters of Tiritiri Matangi
www.tiritirimatangi.org.nz
enquiries@tiritirimatangi.org.nz
Bookings accommodation: www.doc.govt.
nz/tiritiribunkhouse / Tel: 09 425 7812

Wellington Botanical Society
www.wellingtonbotsoc.wellington.net.nz

Whale & Dolphin Adoption Project
adopt.whale.dolphin@clear.net.nz
Tel: 09 521 3999 / Fax: 09 521 1250

WWF New Zealand
www.wwf.org.nz
info@wwf.org.nz
Tel: 0800 4357 993 / Fax: 04 499 2954

Zoological Society of Auckland
www.zoologicalsociety.co.nz

INDEX

PHOTO CREDITS

All photographs author JULIAN FITTER unless credited otherwise below.

ROBIN BUSH Cook's Petrel p 39, New Zealand Storm Petrel p 41, Little Tern, Grey Ternlet, Fairy Tern p 49, Nankeen Kestrel p 67

A. CHAMBERS Barn Owl p 67

RICHARD CHAMBERS Celmisia, Mt Cook p 23, Kauri leaves and cones p 133, Yellow Silver Pine p 135, Pokaka juvenile foliage p 147, Kumerahou p 149, Ribbonwood p 151, Large-leaved Milk Tree p 151, Ramarama p 165, Round-leaved Coprosma p 171, Mount Cook Lily p 205

STUART CHAMBERS Little Egret p 51, Black-fronted Dotterel p 61, Eastern Rosella p 69, Stitchbird male p 73, Brown Quail p 79

MIKE DANZENBAKER Black-bellied Storm Petrel p 41

DEPARTMENT OF CONSERVATION © Crown copyright Grey-backed Storm Petrel [Don Merton 1984] p 41, Marsh Crake [Peter J Moore 1983] p 57, Chukar [Hans Rook 1983] p 79, Kauri Snail [Rod Morris 1980], Flax Snail [Gregory H. Sherley 1996] p 99

SIMON FORDHAM/NATURE PIX Bellbird male p 73, Brown Creeper p 75

DAVID HAWKINS Blackbird p 81, Chaffinch p 83

DON HORNE/NATURAL IMAGE PHOTOGRAPHY Australian Painted Lady Butterfly p 103, Long-tailed Blue Butterfly p 105, Cabbage Tree Moth, Lichen Moth p 107, Large Sand Scarab p 109, Little Grass Cicada p 115, White-footed Ant, Red Percher Dragonfly p 119

ISTOCK Common Bottlenose Dolphin, Southern Right Whale p 85

BRIAN LLOYD Long-tailed Bat, Lesser Short-tailed Bat p 89

RUSSELL MCGEORGE/DAVID BATEMAN LTD Arawhata River p 9, Mt Ruapehu p 15, Rangitikei River p 16, Hooker Valley, Tarn Key Summit p 23, Rakaia River p 24, Lake Taupo, Waihou River,

Cardrona River, Awatere River, Caples River, Lake Pukaki p 25, Lake Rotoroa, Hamilton p 27

IAN MONTGOMERY Australasian bittern p 51

NATURAL SCIENCES IMAGE LIBRARY Maud Island [Gideon Climo] p 11, Banks Peninsula [Landcare Research] p 27, Tui [Peter E. Smith] p 28, p 73, Broad-billed Prion in burrow [Pauline Syrett] p 41, Eastern Rockhopper Penguin [Peter E. Smith] p 43, Black-billed Gull [Peter E. Smith], Black-backed Gull juvenile [Peter E. Smith] p 47, White Heron [Peter E. Smith], Cattle Egret [Peter E. Smith], Reef Heron [Peter E. Smith], Royal Spoonbill [Peter E. Smith] p 51, Cape Barren Goose [Peter E. Smith], Feral or Farmyard Goose [Peter E. Smith], Australian Crested Grebe [Peter E. Smith] p 53, NZ shoveler [Peter E. Smith] p 55, Banded Rail [Peter E. Smith] p 57, Wrybill [Angus McIntosh] p 59, Rock Pigeon [Peter E. Smith], Long-tailed Cuckoo [Landcare Research], Kingfisher [Landcare Research] p 71, Stitchbird female [Peter E. Smith] p 73, Fernbird [Peter Righteous] p 75, Rock Wren male [Landcare Research], Grey Warbler [Peter E. Smith] p 77, Indian peafowl [Peter E. Smith], Helmeted guineafowl [Peter E. Smith] p 79, Magpie [Peter E. Smith], Rook [Peter E. Smith], Song Thrush [Peter E. Smith], Starling [Peter E. Smith] p 81, House Sparrow [Peter E. Smith], Goldfinch [Peter E. Smith] p 83, Bull Elephant Seal [John Marris] p 87, Red-necked Wallaby [G. R. 'Dick' Roberts, Hare [Peter E. Smith], Hedgehog [G. R. 'Dick' Roberts] p 89 Thar [Peter E. Smith], Chamois [Peter E. Smith], Ferret [Peter E. Smith], Whitetail Deer [Landcare Research], Feral Goat [Landcare Research], Red Deer [Peter Righteous] p 91, Archey's Frog [John Marris], Hochstetter's Frog [John Marris], Southern Bell Frog [Peter E. Smith], Whistling Frog [Peter E. Smith] p 93, Common Gecko [Landcare Research], Green Gecko [Landcare Research], Jewelled Gecko [Keven Drew] p 95, Common Skink [Peter E. Smith] p 97, Leaf-veined Slug [John Marris], Peripatus [Peter E. Smith], Giant Land Snail [G. R. 'Dick' Roberts], p 99, Sheetweb Spider, Katipo Spider, Garden

Orbweb Spider [all Peter E. Smith] p 101, Yellow Admiral Butterfly [Geoff Bryant] p 103, Southern Blue Butterfly [G. R. 'Dick' Roberts], Cabbage White Butterfly [Peter E. Smith] p 105, Gum Emperor Moth [John Marris], Kumara Moth [Peter E. Smith], Magpie Moth [Peter Righteous], Silver Y Moth, Cinnabar Moth [G. R. 'Dick' Roberts] p 107, Eleven-spotted Ladybird, Two-Spotted Ladybird, Orange-spotted, Steel-blue Ladybird [all Peter E. Smith] p 109, Earl's Stag Beetle [John Marris], Flax Weevil [John Marris], Speargrass Weevil [Peter E. Smith], Gorse Seed Weevil [Landcare Research], Giraffe Weevil [Landcare Research], Helm's Stag Beetle [John Marris], Metallic Ground Beetle [Peter E. Smith] p 111, Giant Weta [John Marris], Ground Weta [Pauline Syrett], Mountain Stone Weta [Peter E. Smith] p 113, Migratory Locust [Peter E. Smith], Northern Tussock Grasshopper [Peter E. Smith], Southern Tussock Grasshopper [Peter Righteous], Katydid [Peter E. Smith] p 115, Common Wasp, Ichneumon Wasp [both Peter E. Smith] p 117, Black Hunting Wasp, Bush Giant Dragonfly [both Peter E. Smith] p 119, Blue Damselfly [Peter E. Smith], Red Damselfly [Peter E. Smith], Common Stick Insect [G. R. 'Dick' Roberts], Spiny Stick Insect [G. R. 'Dick' Roberts] p 121, Black Cockroach, Flightless Bush Cockroach, New Zealand Praying Mantis [all Peter E. Smith] p 123, Speckled Whelk [Peter E. Smith] p 124, Tuatua, Tuangi [both Peter E. Smith], Pacific Oyster [G. R. 'Dick' Roberts] p 125, Tunneling Mud Crab, Hermit Crab [both Peter E. Smith] p 126, New Zealand Scallop, Horse Mussel, Blackfoot Paua [all Peter E. Smith] p 127, Red Anenome [G. R. 'Djck' Roberts] p 128, Reef Star, Mottled Brittlestar [both Peter E. Smith] p 129, Kawaka [Geoff Bryant] p 133, Miro [Landcare Research] p 133, Pahautea cones p 133, Matai [Peter Smith] p 135, Mountain Toatoa [Landcare Research] p 137, Black Beech (both pics) [G.R. 'Dick' Roberts/Geoff Bryant], Hard Beech trunk [Keven Drew], Red Beech flower [Geoff Bryant] p 139, White Maire [Geoff Bryant] p 145, Toru [Geoff Bryant] p 147, Mountain Mahoe [Geoff Bryant] p 149, Pukatea flowers [G.R. 'Dick' Roberts] p 151, Kaikomako [Landcare Research] p 153, Weeping Matipo [G.R. 'Dick' Roberts] p 167, Willow-leaved Hebe [G. R. 'Dick' Roberts], Koromiko [G.R. 'Dick' Roberts] p 173, Kohia [Landcare Research] p 187, Fork Fern [G.R. 'Dick' Roberts] p 247

MONIQUE NELSON-TUNLEY Shore Skink p 93

STEVE REEKIE Giant stick insect p 121

JEREMY ROLFE Pink Pine (both) p 137, Pokaka p 147, Horopito, Wavy-leaved Coprosma p 163, Mamangi p 165, Turepo p 167, Bush Snowberry,
Tauhini p 169, Coprosma crassifolia, Red-fruited Karamu p 171, Tree Hebe p 173, Earina mucronata p 189, Giant Speargrass p 203, Sand Tussock p 219, Creeping Tree Fern p 229, Long-leaved Fork Fern p 247

ROVING TORTOISE PHOTOGRAPHY/TUI DE ROY Okarito Brown Kiwi p 4, Southern Royal Albatross p 31, Black-browed Mollymawk (both pics), Grey-headed Mollymawk, Light-mantled Albatross, Salvin's Mollymawk, Shy Mollymawk (both pics) p 33, Northern Giant Petrel (in flight) p 35, Common Diving Petrel p 41, Yellow-eyed Penguin, Little Blue Penguin, Snares Crested Penguin, Fiordland Crested Penguin p 43, Little Shag p 45, Grey Teal p 55, Black Stilt p 59, Banded Dotterel p 61, All p 65, New Zealand Falcon p 67, Kakapo, Kaka, Yellow-crowned Parakeet p 69, Yellowhead p 75, Rifleman male p 77, Sperm Whale, Humpback Whale p 85, Brush-tailed Possum p 89, Green and Golden Bell Frog p 93, Black Tunnelweb Spider p 101, Tree Weta, Cave Weta p 113

ROVING TORTOISE PHOTOGRAPHY/MARK JONES Orca p 85, Hector's Dolphin p 87, New Zealand Glow Worm p 123

J. SHARKEY Greenfinch p 83

JOHN SMITH-DODSWORTH Toatoa p 137, Tawheowheo p 153, Hutu p 155, Snow Gentian, Common Gentian p 205, Penwiper p 209, Silver Cushion Daisy p 215, Punui p 229, Lyall's Spleenwort p 239

R VALENTINO/DAVID BATEMAN LTD Tarawera River p 25, all page 27 except Lake Rotoroa, Hamilton and Banks Peninsula

TONY WHITAKER Hamilton's Frog, Green and Gold Bell Frog p 93, Forest Gecko p 95, Rainbow Skink, Ornate Skink, Green Skink, Copper Skink, Scree Skink p 97

STEVE WOOD All except Northern Giant Petrel p 35, All except White-chinned Petrel p 37, Black-winged Petrel, Soft-plumaged Petrel, Hutton's Shearwater, Little Shearwater, Fairy Prion p 39, Broad-billed Prion p 41, Little Black Shag p 45, Arctic Skua p 47, Black-fronted Tern p 49, Grey Duck, Mallard p 55, Spotless Crake p 57, Pied Oystercatcher p 59, Pacific Golden Plover, Bar-tailed Godwit inset p 61, All p 63, Australasian Harrier (both), Little Owl p 67, Welcome Swallow, Shining Cuckoo p 71. Whitehead male p 75, Rifleman female p 77, Common Pheasant p 79, Skylark p 81, Cirl Bunting, Dunnock, Lesser Redpoll p 83